PLANAR GRAPHS
THEORY AND ALGORITHMS

T. NISHIZEKI
Tohoku University
Sendai, Japan

N. CHIBA
Iwate University
Morioka, Japan

DOVER PUBLICATIONS, INC.
Mineola, New York

Bibliographical Note

This Dover edition, first published in 2008, is an unabridged republication of the work originally published as Volume 32 of the North-Holland Series *Annals of Discrete Mathematics* by North-Holland, Amsterdam and New York, in 1988.

Library of Congress Cataloging-in-Publication Data

Nishizeki, T. (Takao), 1947–
    Planar graphs : theory and algorithms / T. Nishizeki, N. Chiba.—Dover ed.
        p. cm.
    "This Dover edition, first published in 2008, is an unabridged republication of the work originally published as Volume 32 of the North-Holland Series Annals of Discrete Mathematics by North-Holland, Amsterdam and New York, in 1988."
    Includes bibliographical references and index.
    ISBN-13: 978-0-486-46671-2
    ISBN-10: 0-486-46671-X
    1. Graph theory. 2. Algorithms. I. Chiba, N. (Norishige) II. Title.

QA166 .N57 2008
511'.5—dc22
                                                    2008005295

Manufactured in the United States of America
Dover Publications, Inc., 31 East 2nd Street, Mineola, N.Y. 11501

CONTENTS

Dedicated to our teacher
Professor Nobuji Saito

ACKNOWLEDGMENTS

It is a pleasure to record the authors' gratitude to those to whom we are indebted, directly or indirectly, in writing this book. First of all, our thanks are due to Professor Nobuji Saito of Tohoku University under whose guidance we have studied and worked since our student days; without his continual encouragement this book would not have been completed.

The present book is based primarily upon a series of the authors' own investigations. We wish to thank the coauthors of our joint papers: S. Abe, T. Asano, I. Baybars, H. Gabow, D. Hochbaum, O. Kariv, D. Leven, K. Matsumoto, K. Onoguchi, T. Ozawa, M. Sato, D. Shmoys, K. Suzuki, H. Suzuki, K. Takamizawa, O. Terada, T. Watanabe and T. Yamanouchi.

A book such as this one owes a great deal, of course, to many previous workers and writers. Without trying to be complete, we would like to mention the very important succession of books and papers by Aho, Hopcroft and Ullman; Baker; Behzad, Chartrand and Lesniak-Foster; Bollobás; Booth and Lueker; Even; Frederickson; Golumbic; Hassin; Hopcroft and Tarjan; Lipton and Tarjan; Miller; Okamura; Okamura and Seymour; Ore; Papadimitriou and Steiglitz; Seymour; Tarjan; Thomassen; Tutte; and Wilson.

Finally, following a Western custom, we should thank our better halves, Yuko Nishizeki and Yumiko Chiba for many things which have nothing at all to do with planar graphs.

PREFACE

The theory of planar graphs was first discovered in 1736 by Euler when he found his important formula relating the numbers of vertices, edges and faces of polyhedrons, which can be represented by planar graphs. Since that time numerous results have been obtained on planar graphs. One of the most outstanding results is Kuratowski's theorem which gives a criterion for a graph to be planar. Another example is the famous four-color theorem: every planar graph can be colored with at most four colors so that no two adjacent vertices receive the same color. In recent years, planar graphs have attracted computer scientists' interest, and a lot of interesting algorithms and complexity results have been obtained for planar graphs. For example, Hopcroft and Tarjan have reported on a linear time algorithm which tests the planarity of a graph.

Recently it appeared to us that the time was ripe to collect and organize the many results on planar graphs, which have been our research topics for these ten years. In our opinion the theory and algorithms are complementary to each other in the research of planar graphs. For example, Hopcroft and Tarjan's algorithm was motivated by Kuratowski's theorem although it was not explicitly used in the algorithm. On the other hand many theoretic results have been obtained from the algorithmic view point. Thus we have tried to include most of the important theorems and algorithms that are currently known for planar graphs. Furthermore we have tried to provide constructive proofs for theorems, from which algorithms immediately follow. Most of the algorithms are written in Pidgin PASCAL in a manner that will make their adaptation to a practical programming language relatively easy. They are all efficient, and most of them are the best known ones; the complexities are linear or $O(n \log n)$.

A glance at the table of contents will provide an outline of the topics to be discussed. The first two chapters are introductory in the sense that they provide the foundations, respectively, of the graph theoretic notions and algorithmic techniques that will be used in the book. Experts in graph theory or algorithms may skip Chapters 1 or 2. The remaining chapters discuss the topics on planarity testing, embedding, drawing, vertex- or edge-coloring, maximum independent set, subgraph listing, planar separator theorem, Hamiltonian cycles, and single- or multicommodity flows. The topics reflect the authors' favor as graph theorists and computer scientists. The chapters are structured in such a way that the book will be suitable as a textbook in a course on algorithms, graph theory, or planar graphs. In addition, the book will be useful for computer scientists and graph theorists at the research level. An extensive reference section is also included.

Sendai   Takao Nishizeki and Norishige Chiba

CHAPTER 1

GRAPH THEORETIC FOUNDATIONS

1.1. Introduction

A graph can be thought of as a diagram consisting of a collection of vertices together with edges joining certain pairs of vertices. A planar graph is a particular diagram which can be drawn on the plane so that no two edges intersect geometrically except at a vertex at which they are both incident.

First consider the example depicted in Fig. 1.1(a), which consists of six vertices (drawn by small black cycles) and 12 edges (drawn by straight lines there). Is the graph planar? That is, can you draw the graph on the plane by locating vertices and drawing edges appropriately in such a way that no two edges intersect except at a common endvertex? The drawing in Fig. 1.1(a), as it is, has two intersections in the circles drawn by dotted lines. However, one can avoid them if the vertex v_6 is located in the exterior of the square $v_1v_2v_3v_4\ v_1$ (drawn by a thick line), as shown in Fig. 1.1(b). Thus the graph is known to be planar. Next consider the graph depicted in Fig. 1.2(a), known as the complete graph K_5 on five vertices. Is K_5 planar? If we suppose so, then one may assume without loss of generality that $v_1v_2v_3v_4v_5$ is drawn on the plane as a regular pentagon. (Look on the plane as flexible rubber, and deform it as desired.) One may also assume that the edge (v_1, v_3) is drawn in the interior of the pentagon. Then both the edges (v_2, v_5) and (v_2, v_4) must be drawn in the exterior, and consequently edge (v_3, v_5) must be drawn in the interior, as shown in Fig. 1.3(a). Then an intersection must occur whether the edge (v_1, v_4) is drawn in the interior or exterior. Thus K_5 cannot be drawn on the plane without edge-crossing, so is nonplanar. Another example of nonplanar graphs is the "complete bipartite graph" $K_{3,3}$ depicted in Fig. 1.2(b). One may assume that edge (u_1, v_2) is drawn in the interior of the hexagon $u_1v_1u_2v_2u_3v_3$, and hence edge (v_1, u_3) in the exterior. Then (u_2, v_3) cannot be drawn without producing an intersection. Thus $K_{3,3}$ is also known to be nonplanar.

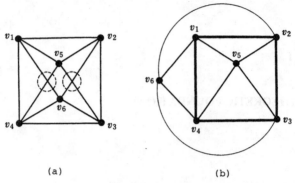

Fig. 1.1. (a) A graph G; and (b) A plane embedding of G.

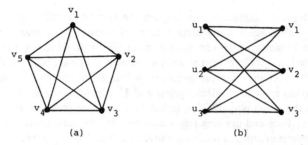

Fig. 1.2. Kuratowski's graphs: (a) Complete graph K_5; (b) Complete bipartite graph $K_{3,3}$.

As above, not all graphs are planar. However planar graphs arise quite naturally in real-world applications, such as road or railway maps, electric printed circuits, chemical molecules, etc. Planar graphs play an important role in these problems, partly due to the fact that some practical problems can be efficiently solved for planar graphs even if they are intractable for general graphs. Moreover, a number of interesting and applicable results are known concerning the mathematical and algorithmic properties of planar graphs. Thus the theory of planar graphs has emerged as a worthwhile mathematical discipline in its own right.

1.2. Some basic definitions

Let us formally define the notion of a graph. A *graph* $G = (V, E)$ is a structure which consists of a finite set of *vertices* V and a finite set of *edges* E;

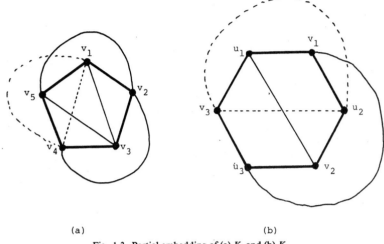

Fig. 1.3. Partial embedding of (a) K_5 and (b) $K_{3,3}$.

each edge is an unordered pair of distinct vertices. Throughout the book n denotes the number of vertices of G, that is, $n = |V|$, while m denotes the number of edges, that is, $m = |E|$. Any edge of the form (u, v) is said to *join* the vertices u and v. Our graph G is a so-called *simple finite graph*, that is, G has no "multiple edges" or "loops" and sets V and E are finite. *Multiple edges* join the same pair of vertices, while a *loop* joins a vertex to itself. The graph, in which loops and multiple edges are allowed, is called a *multigraph*. The graph, in which E is defined to be a set of ordered pairs of distinct vertices, is a *directed graph* (*digraph* for short).

If $(u, v) \in E$, then two vertices u and v of a graph G are said to be *adjacent*; u and v are then said to be *incident* to edge (u, v); u is a *neighbour* of v. The *neighbourhood* $N(v)$ is the set of all neighbours of v. Two distinct edges are *adjacent* if they have a vertex in common. The *degree* of a vertex v of G is the number of edges incident to v, and is written as $d(G, v)$ or simply $d(v)$. In the graph G depicted in Fig. 1.1(a) vertices v_1 and v_2 are adjacent; $N(v_1) = \{v_2, v_4, v_5, v_6\}$, and hence $d(v_1) = 4$.

We say that $G' = (V', E')$ is a *subgraph* of $G = (V, E)$ if $V' \subset V$ and $E' \subset E$. If $V' = V$ then G' is called a *spanning* subgraph of G. If G' contains all the edges of G that join two vertices in V' then G' is said to be *induced by* V'. If V' consists of exactly the vertices on which edges in E' are incident, then G' is said to be *induced by* E'. Fig. 1.4(a) depicts a spanning subgraph of G in Fig. 1.1(b); Fig. 1.4(b) depicts a subgraph induced by $V' = \{v_1, v_2, v_4, v_5\}$; Fig. 1.4(c) depicts a subgraph induced by $\{(v_1, v_2), (v_1, v_4), (v_1, v_5), (v_2, v_5)\}$.

We shall often construct new graphs from old ones by deleting some vertices or edges. If $V' \subset V$ then $G - V'$ is the subgraph of G obtained by deleting the vertices in V' and all edges incident on them, that is, $G - V'$ is a subgraph induced by $V - V'$. Similarly if $E' \subset E$ then $G - E' = (V, E - E')$. If $V' = \{v\}$ and $E' = \{(u, v)\}$ then this notation is simplified to $G - v$ and $G - (u, v)$.

We also denote by G/e the graph obtained by taking an edge e and *contracting* it, that is, removing e and identifying its ends u and v in such a way that the resulting vertex is adjacent to those vertices (other than u and v) which were originally adjacent to u or v. For $E' \subset E$ we denote by G/E' the graph which results from G after a succession of such contractions for the edges in E'. The graph G/E' is called a *contraction* of G.

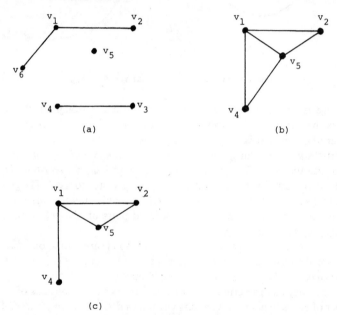

Fig. 1.4. Subgraphs of G in Fig. 1.1(b): (a) Spanning subgraph; (b) Vertex-induced subgraph; (c) Edge-induced subgraph.

A v_0-v_l *walk* in G is an alternating sequence of vertices and edges of G, $v_0, e_1, v_1, \ldots, v_{l-1}, e_l, v_l$, beginning and ending with a vertex, in which each edge is incident on two vertices immediately preceding and following it. The number l of edges is called its *length*. If the vertices $v_0, v_1, \ldots, v_l$ are distinct (except, possibly, $v_0 = v_l$), then the walk is called a *path* and is usually denoted by $v_0 v_1 \cdots v_l$. A path or walk is *closed* if $v_0 = v_l$. A closed path containing at least one edge is called a *cycle*. A cycle of length $3, 4, 5, \ldots$, is called a triangle, quadrilateral, pentagon, etc. One example of walks in G depicted in Fig. 1.1(b) is

$$v_1, (v_1, v_2), v_2, (v_2, v_3), v_3, (v_3, v_5), v_5,$$

$$(v_5, v_2), v_2, (v_2, v_3), v_3, (v_3, v_4), v_4,$$

which is not closed, that is *open*. One example of cycles is $v_1 v_2 v_3 v_4 v_1$, a quadrilateral.

The dual concept of a cycle is a "cutset" which we now define. A *cut* of a graph G is a set of edges of G whose removal increases the number of components. A *cutset* is defined to be a cut no proper subset of which is a cut, that is, a cutset is a minimal cut. Fig. 1.5 illustrates these concepts; $\{a, b, c, d, e\}$ is a cut but not a cutset; both $\{a, b, c\}$ and $\{d, e\}$ are cutsets.

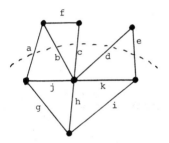

Fig. 1.5. Cut and cutset.

A graph G is *connected* if for every pair $\{u, v\}$ of distinct vertices there is a path between u and v. A (*connected*) *component* of a graph is a *maximal connected subgraph*. A *cutvertex* is a vertex whose deletion increases the number of components. Similarly an edge is a *bridge* if its deletion increases the number of components. G is *2-connected* if G is connected and has no cutvertex. A *block* of G is a maximal 2-connected subgraph of G. A *separation pair* of a 2-connected graph G is two vertices whose deletion disconnects G. G is *3-connected* if G has no cutvertex or separation pair. In general, a *separating set* of a connected graph G is a set of vertices of G whose deletion disconnects G. The graph in Fig. 1.4(a) is disconnected, and has three components; the graph in Fig. 1.4(c) which is not 2-connected has a cutvertex v_1, a bridge (v_1, v_4) and two blocks; the graph in Fig. 1.4(b) has no cutvertex but has a separation pair $\{v_1, v_5\}$, so is 2-connected but not 3-connected; G in Fig. 1.1(b) has no cutvertex or separation pair, so is 3-connected.

If G has a separation pair $\{x, y\}$, then we often split G into two graphs G_1

and G_2, called *split graphs*. Let $G_1' = (V_1, E_1')$ and $G_2' = (V_2, E_2')$ be two sub-graphs satisfying the following conditions (a) and (b):

(a) $V = V_1 \cup V_2$, $V_1 \cap V_2 = \{x, y\}$;

(b) $E = E_1' \cup E_2'$, $E_1' \cap E_2' = \varnothing$, $|E_1'| \geqslant 2$, $|E_2'| \geqslant 2$.

Define G_1 to be the graph obtained from G_1' by adding a new edge (x, y) if it does not exist; similarly define G_2. (See Fig. 1.6.)

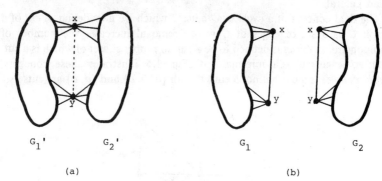

Fig. 1.6. (a) A graph G with a separation pair $\{x, y\}$, where edge (x, y)
may not exist; (b) Split graphs G_1 and G_2.

Before ending this section, we will define some special graphs. A graph without any cycles is a *forest*; a *tree* is a connected forest. Fig. 1.4(a) is a forest having three components.

A graph in which every pair of distinct vertices are adjacent is called a *complete graph*. The complete graph on n vertices is denoted by K_n. K_5 has been depicted in Fig. 1.2(a).

Suppose that the vertex set V of a graph G can be partitioned into two disjoint sets V_1 and V_2, in such a way that every edge of G joins a vertex of V_1 to a vertex of V_2; G is then said to be a *bipartite graph*. If every vertex of V_1 is joined to every vertex of V_2, then G is called a *complete bipartite graph* and is denoted by $K_{s,r}$ where $s = |V_1|$ and $r = |V_2|$. Fig. 1.2(b) depicts a complete bipartite graph $K_{3,3}$ with partite sets $\{u_1, u_2, u_3\}$ and $\{v_1, v_2, v_3\}$.

1.3. Planar graphs

Let us formally define a planar graph. Draw a graph G in the given space (e.g. plane) with points representing vertices of G and curves representing edges. G

can be *embedded* (or has an *embedding*) in the space if G can be drawn in such a way that no two edges intersect except at an endvertex in common. A graph G is *planar* if G has an embedding in the plane. Considering the stereographic projection depicted in Fig. 1.7, one can easily establish that a graph is planar if and only if it can be embedded on the surface of a sphere. A *plane graph* is a planar graph which is embedded in the plane. Thus Fig. 1.1(a) is a planar graph, while Fig. 1.1(b) is a plane graph.

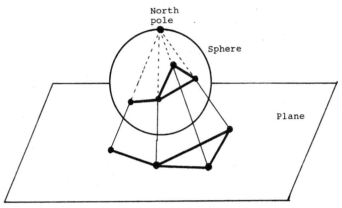

Fig. 1.7. Stereographic projection.

Delete from the plane all curves and points corresponding, respectively, to the edges and vertices of a plane graph G, then the remainder fall into connected components, called *faces*. Note that one face is unbounded; it is called an *outer* (or *infinite*) *face*. If G is embedded on the sphere, then no face of G can be regarded as an outer one. On the other hand, clearly a plane graph G can always be embedded in the plane so that a given face of G becomes the outer face. (Consider again the stereographic projection in Fig. 1.7, and rotate the sphere so that the north pole is in that face.) The *boundary* of a face F is the set of edges in the closure of the face. Thus the boundary is a walk in general, and is a cycle if G is a 2-connected graph with at least three vertices. A cycle in a plane graph G is called a *facial cycle* if it is a boundary of a face.

In general a planar graph has many embeddings in the plane. We shall now define an equivalence relation among these embeddings. Two embeddings of a planar graph are *equivalent* when the boundary of a face in one embedding always corresponds to the boundary of a face in the other. If G is a disconnected plane graph, one can obtain a new nonequivalent embedding simply by replacing a connected component within another face. Similarly, if G has a cutvertex v, one may obtain a new nonequivalent embedding by replacing a component of $G - v$ (together with the edges joining v and vertices in the component) in another face incident to v. Thus we shall assume that G is 2-connected if the embedding is unique. We say that the plane embedding of a graph is *unique* when the embeddings are all equivalent. Whitney [Whi33b] proved that the embedding of a 3-connected planar graph is unique. Before proving the result, we need two more terms. A graph G' is said to be a *subdivision* of a graph G if G' is obtained from G by subdividing some of the edges, that is, by replacing the edges by paths having at most their endvertices in common. Fig. 1.8 depicts subdivisions of K_5 and $K_{3,3}$. If C is a cycle of G, then a *C-component* (or *bridge*) of G is either an edge (together with its ends) not in C joining two vertices of C or it is a connected component of $G - V(C)$ together with all edges (and their ends) of G joining this component to C. Fig. 1.9 illustrates a cycle C and two C-components H and H'.

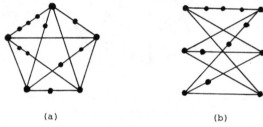

(a)   (b)

Fig. 1.8. Subdivisions of (a) K_5 and (b) $K_{3,3}$.

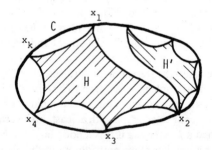

Fig. 1.9. Facial cycle C with two C-components H and H'.

Theorem 1.1. *The embedding of a 2-conneted planar graph G is unique if and only if G is a subdivision of a 3-connected graph.*

Proof. *Necessity*: Suppose that a 2-connected planar graph G is not a subdivision of a 3-connected graph. Then there is a separation pair having split graphs G_1 and G_2 such that both G_1' and G_2' are not paths. (See Fig. 1.6.) Then a new embedding of G is obtained by a reflection or twist of G_1' or G_2'. The boundary of the outer face in the original embedding is no longer a face boundary in the new embedding. Thus the embedding is not unique.

Sufficiency: Suppose that the embedding of a 2-connected planar graph G is not unique. Thus, according to the definition, the original embedding $Em(G)$ of G has a face F with the facial cycle C such that C is no longer a facial cycle in another embedding $Em'(G)$ of G. Clearly G has two C-components H and H'; one in the interior and the other in the exterior of C in $Em'(G)$. One may assume that C is the boundary of the outer face in $Em(G)$. Let $x_1, x_2, \ldots, x_k$ be the vertices of C contained in H occurring in cyclic order. One may assume that all the vertices of C contained in H' are in the subpath of C joining x_1 and x_2 and containing no other x_i. (See Fig. 1.9.) Then $\{x_1, x_2\}$ is a separation pair, for which both G_1' and G_2' are not paths. Therefore G is not a subdivision of a 3-connected graph. Q.E.D.

Theorem 1.1 immediately implies that every 3-connected planar graph has a unique plane embedding.

1.4. Euler's formula

There is a simple formula relating the numbers of vertices, edges and faces in a connected plane graph. It is known as Euler's formula because Euler established it for those plane graphs defined by the vertices and edges of polyhedra. In this section we discuss Euler's formula and its immediate consequences.

Theorem 1.2. (*Euler 1750*). *Let G be a connected plane graph, and let n, m, and f denote respectively the number of vertices, edges and faces of G. Then $n - m + f = 2$.*

Proof. We employ an induction on m, the result being obvious for $m = 0$ or 1. Assume that the result is true for all connected plane graphs having fewer than m edges, where $m \geqslant 2$, and suppose G has m edges. Consider first the case G is a tree. Then G has a vertex v of degree one. The connected plane graph $G - v$ has $n - 1$ vertices, $m - 1$ edges and f faces, so that by the inductive hypothesis, $(n - 1) - (m - 1) + f = 2$, which implies that $n - m + f = 2$. Consider next the case when G is not a tree. Then G has an edge e on a cycle. In this case the connected plane graph $G - e$ has n vertices, $m - 1$ edges, and $f - 1$ faces, so the desired formula immediately follows from the inductive hypothesis.

Q.E.D.

A *maximal planar* graph is one to which no edge can be added without losing planarity. Thus in any embedding of a maximal planar graph G with $n \geq 3$, the boundary of every face of G is a triangle. Although a general graph may have up to $n(n - 1)/2$ edges, it is not true for planar graphs.

Corollary 1.1. *If G is a planar graph with n (≥ 3) vertices and m edges, then $m \leq 3n - 6$. Moreover the equality holds if G is maximal planar.*

Proof. We can assume without loss of generality that G is a maximal planar graph; otherwise add new edges without increasing n so that the resulting graph is maximal planar. Consider a plane embedding of G. Every face is bounded by exactly three edges, and each edge is on the boundaries of two faces. Therefore, counting up the edges around each face, we have $3f = 2m$. Applying theorem 1.2, we obtain $m = 3n - 6$.   Q.E.D.

A stronger bound can be obtained for the number of edges of bipartite planar graphs.

Corollary 1.2. *If G is a planar bipartite graph with n (≥ 3) vertices, and m edges, then $m \leq 2n - 4$.*

Proof. Noting that $4f \leq 2m$ and applying Theorem 1.2, we have the desired equation.   Q.E.D.

Corollaries 1.1 and 1.2 immediately imply that K_5 and $K_{3,3}$ are nonplanar. The following result is often useful.

Corollary 1.3. *Let G be a planar graph with n (≥ 3) vertices. Let d_M be the maximum degree of a vertex, and let n_i denote the number of vertices having degree i, where $i = 1, 2, \ldots, d_M$. Then*

$$5n_1 + 4n_2 + 3n_3 + 2n_4 + n_5 \geq n_7 + 2n_8 + \cdots + (d_M - 6)n_{d_M} + 12.$$

Proof. Clearly $n = \Sigma n_i$ and $2m = \Sigma in_i$ where summations are over $i = 1, \ldots, d_M$. Since $6n \geq 2m + 12$ by Corollary 1.1, it follows that

$$6 \Sigma n_i \geq \Sigma in_i + 12,$$

from which the desired inequality immediately follows.   Q.E.D.

The following corollary immediately follows from the corollary above.

Corollary 1.4. *Every planar graph contains a vertex of degree at most 5.*

1.5. Kuratowski's theorem

One of the most beautiful theorems in graph theory is Kuratowski's, which gives a characterization of planar graphs in terms of "forbidden graphs". In this section we present and prove the theorem.

Since both K_5 and $K_{3,3}$ are nonplanar as seen twice in Sections 1.1 and 1.4, every planar graph does not contain a subdivision of K_5 and $K_{3,3}$ depicted in Fig. 1.8. Surprisingly the converse is also true.

Theorem 1.3. (*Kuratowski* 1930) *A graph is planar if and only if it does not contain a subdivision of K_5 and $K_{3,3}$.*

Since we have already seen the necessity of Theorem 1.3, we shall prove only the sufficiency. The following proof is adapted from [Tho80]. First we give the following lemmas.

Lemma 1.1. [*Tho*80] *If G is a 3-connected graph having five or more vertices, then G contains an edge e such that the graph G/e obtained from G by contracting e is 3-connected.*

Proof. Suppose that the claim is not true. Then G/e is not 3-connected for any edge $e = (x, y)$ of G. Since G/e has no cutvertex, it must have a separation pair of vertices, one of which is the vertex obtained by identifying x and y. Thus G has a separation set of the form $\{x, y, z\}$. Assume that e and z are chosen in a way that the largest component H of $G - \{x, y, z\}$ is largest possible. Let H' be another component of $G - \{x, y, z\}$, and let $e' = (z, u)$ be an edge joining z and a vertex u in H'. Since G/e' is not 3-connected, G has a separating set of the form $\{z, u, v\}$. (See Fig. 1.10.) Clearly either vertex v is in H' or $v = x$ or y. If v is in H', then clearly $G - \{z, u, v\}$ has a component including all the vertices in H together with x and y, contradicting the assumption about the maximality of H. Thus $v = x$ or y, and we may assume $v = x$. Then the subgraph $H + y$ of G induced by the vertices of H together with y must be connected; for, otherwise, $\{z, x\}$ would be a separation pair of G. Therefore $H + y$ is contained in a single component of $G - \{z, u, v\}$, contradicting the assumption about the maximality of H. Q.E.D.

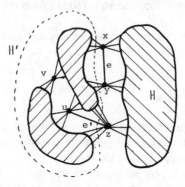

Fig. 1.10. Illustration for proof of Lemma 1.1.

Lemma 1.2. *Let e be an edge in a graph G. If G/e contains a subdivision of K_5 or $K_{3,3}$, then G also contains a subdivision of K_5 or $K_{3,3}$.*

Proof. Let $e = (x, y)$ and let z be the vertex obtained by identifying x and y. Let S be the subdivision of K_5 or $K_{3,3}$ contained in G/e. Each edge (z, v) incident on z in G/e may have two counterparts in G: edges (x, v) and (y, v). Arbitrarily choosing one of them for each edge incident on z, construct a subgraph T of G induced by the edges in S together with e. If z has degree 2 in S, then clearly T and hence G contain a subdivision of K_5 or $K_{3,3}$. Thus we may assume that z has degree 3 or more. Furthermore if x or y has degree 2 in T, then clearly T and hence G contain a subdivision of K_5 or $K_{3,3}$. Hence S must be a subdivision of K_5 and both x and y must have degree 3 in T as shown in Fig. 1.11. As indicated by thick lines, T and hence G contain a subdivision of $K_{3,3}$.   Q.E.D.

We are now ready to prove the sufficiency of Theorem 1.3.

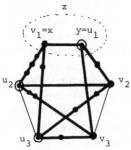

Fig. 1.11. Subgraph T in G, which contains a subdivision of $K_{3,3}$ denoted by thick lines.

Sufficiency of Theorem 1.3. We prove by induction on n the sufficiency: G is planar if G does not contain a subdivision of K_5 and $K_{3,3}$. Since $K_5 - e$ is planar for any edge in K_5, the statement is true if $n \leq 5$. We now assume that G has $n \geq 6$ vertices and that the statement is true for graphs having fewer than n vertices. There are two cases to consider.

Case 1: G is not 3-connected. Clearly a graph is planar if and only if each of its blocks is planar, so we may assume that G is 2-connected. Therefore G has a separation pair $\{x, y\}$. Let G_1 and G_2 be the split graphs with respect to the pair. (See Fig. 1.6.) Clearly both G_1 and G_2 contain fewer vertices than G, and contain no subdivision of K_5, or $K_{3,3}$. Therefore by inductive hypothesis both G_1 and G_2 are planar. Furthermore both have plane embeddings such that edge (x, y) lies on the outer face boundary. These two plane embeddings can easily be fitted together at x and y to give a plane embedding of G. That is, G is planar.

Case 2: G is 3-connected. Lemma 1.1 implies that G has an edge $e = (x, y)$ such that G/e is 3-connected. Let z be the vertex obtained by identifying x and y. By Lemma 1.2 G/e has no subdivision of K_5 or $K_{3,3}$, and hence by the inductive hypothesis G/e is planar. Consider a plane graph G/e and the plane subgraph $G/e - z$. Since G/e is 3-connected, by Theorem 1.1 the plane embedding of G/e is unique. Let F be the face of the plane graph $G/e - z$ which contained point z, and let C be the boundary cycle of F. Clearly all the neighbours of x or y except themselves must be on cycle C. Let $x_1, x_2, \ldots, x_k$ be the neighbours of x occurring on C in cyclic order and let P_i be the subpath of C joining x_i and x_{i+1} and not containing any x_j, $j \neq i$, $i + 1$, where $x_{k+1} = x_1$. (See Fig. 1.12.) If all neighbours of y except x are contained in one of the paths, then a plane embedding of G can be easily obtained from that of G/e, that is, G is planar, as indicated by dashed lines in Fig. 1.12. Thus not all the neighbours of y except x are contained in one of the paths. Then, since y has three or more neighbours including x, there are three possibilities:
  (a) y has three or more neighbours of $\{x_1, \ldots, x_k\}$;
  (b) y has a neighbour u in $P_i - \{x_i, x_{i+1}\}$ for some i and a neighbour v not in P_i;
  (c) y has two neighbours x_i and x_j such that $j \neq i + 1$ and $i \neq j + 1$.
In case (a) the subgraph of G induced by the vertices in C together with x and y contains a subdivision of K_5. On the other hand, in cases (b) and (c), the subgraph of G induced by the vertices in C together with x and y contains a subdivision of $K_{3,3}$. (See Fig. 1.13.) Thus a contradiction occurs in either case. This completes the proof. Q.E.D.

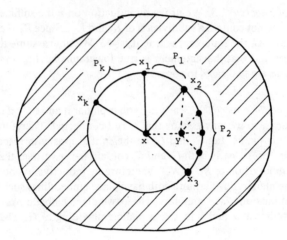

Fig. 1.12. Plane graph $G - y$.

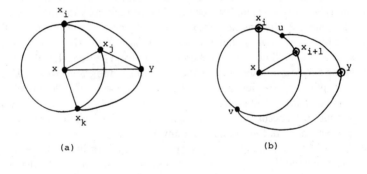

(a)   (b)

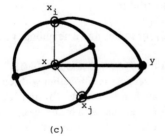

(c)

Fig. 1.13. Subdivisions of K_5 and $K_{3,3}$.

1.6. Dual graphs

Several criteria for planarity have been discovered since the original work of Kuratowski. In this section we first define two types of duals of a graph, i.e., geometric and combinatorial duals, and then give two criteria for planarity in terms of combinatorial duals and in terms of the basis of the "cycle space", respectively. Throughout this section "graphs" means multigraphs.

For a plane graph G, we often construct another graph G^* called the (*geometric*) *dual* of G as follows. A vertex v_i^* is placed in each face F_i of G; these are the vertices of G^*. Corresponding to each edge e of G we draw an edge e^* which crosses e (but no other edge of G) and joins the vertices v_i^* which lie in the faces F_i adjoining e; these are the edges of G^*. The construction is illustrated in Fig. 1.14; the vertices v_i^* are represented by small white circles, the edges e^* of G^* by dashed lines. Clearly G^* has a loop if and only if G has a bridge, and G^* has multiple edges if and only if two faces of G have at least two edges in common. Thus G^* is not necessarily a simple graph even if G is simple. Clearly the geometric dual G^* of a plane graph G is also plane. One can easily observe the following lemma.

Lemma 1.3. *Let G be a plane connected graph with n vertices, m edges and f faces, and let the geometric dual G^* have n^* vertices, m^* edges and f^* faces; then $n^* = f$, $m^* = m$, and $f^* = n$.*

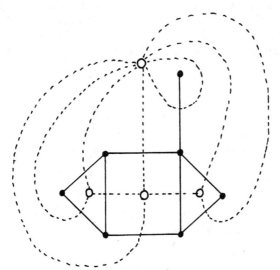

Fig. 1.14. A plane graph G and its geometric dual G^*.

Clearly the dual of the dual of a plane graph G is the original graph G. However a planar graph may give rise to two or more geometric duals since the plane embedding is not necessarily unique. If G is a subdivision of a 3-connected planar graph, then by Theorem 1.1 the plane embedding of G is essentially unique and hence the dual is unique.

The following observation is often useful in designing an efficient algorithm for planar graphs.

Lemma 1.4. *Let G be a planar graph and G^* be a geometric dual of G, then a set of edges in G forms a cycle (or cutset) in G if and only if the corresponding set of edges of G^* forms a cutset (res. cycle) in G^*.*

Proof. One may assume without loss of generality that G is a connected plane graph. Let C be a cycle in G, and let S be the set of vertices of G^* lying in the interior of C. Clearly deleting the edges of G^* corresponding to the edges of C disconnects G^* into two connected subgraphs inside and outside of C: $G^* - S$ and $G^* - (V - S)$. Thus those edges in G^* forms a cutset of G. The remaining claims are similar, and are left to the reader.   Q.E.D.

The property in Lemma 1.4 motivates the following abstract definition of duality. We say that a graph G^* is a *combinatorial dual* of a graph G if there is a one–one correspondence between the edges of G and those of G^* with the property that a set of edges of G forms a cycle if and only if the corresponding set of edges of G^* forms a cutset in G^*. Fig. 1.15 depicts a graph G and its combinatorial dual G^*, with corresponding edges sharing the same letter. If G is a planar graph and G^* is a geometric dual of G, then Lemma 1.4 implies that

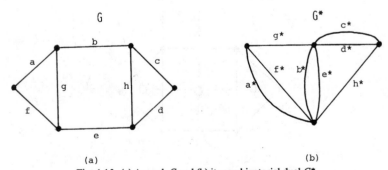

(a)   (b)

Fig. 1.15. (a) A graph G and (b) its combinatorial dual G^*.

G^* is a combinatorial dual. Thus a planar graph necessarily has a combinatorial dual. Whitney [Whi33a] proved that the converse is also true, giving the following criterion for planarity. From now on in this section we assume for simplicity that G is 2-connected.

Theorem 1.4. *A 2-connected graph G is planar if and only if G has a combinatorial dual.*

We have seen the necessity of Theorem 1.4. Before proving the sufficiency, we present and prove another criterion due to MacLane [Mac37], from which the sufficiency of Theorem 1.4 follows rather immediately.

Clearly the set of subsets of a given set E forms a vector space over the field of order 2, where the modulo 2 sum $A \oplus B$ of two subsets A and B of E is defined as the set of all edges that lie either in A or in B but not both, that is, $A \oplus B = (A - B) \cup (B - A)$. If E is the set of edges in a graph G, we call the subspace generated by the cycles of G, the *cycle space* of G. It is easy to verify that if Z and Z' are cycles in G, the sum $Z \oplus Z'$ is an edge-disjoint union of cycles in G. Thus the cycle space of G consists of the cycles of G and the edge-disjoint unions of cycles. A *cycle basis* of G is defined as a basis for the cycle space of G which consists entirely of cycles. A cycle basis B of G is called a *2-basis* of G if every edge of G is contained in at most two cycles of B. The cutset space of G is similarly defined.

It is easily shown that every planar 2-connected graph has a 2-basis. Let B be the collection of all facial cycles except the boundary of the outer face. Every cycle Z is generated by cycles in B: Z is a sum of exactly the facial cycles in the interior of Z. Furthermore the cycles in B are independent: any sum of cycles in B is nonzero. Thus B is a cycle basis of G. An edge in the boundary of the outer face is contained in precisely one cycle in B, while each of the other edges is in two of the facial cycles in B. Hence B is a 2-basis of G. Surprisingly the converse is also true; the result is known as the *McLane's criterion* for planarity. In order to prove the converse we need the following lemma.

Lemma 1.5. *Neither K_5 nor $K_{3,3}$ has a 2-basis.*

Proof. Suppose that K_5 has a 2-basis B. Since the rank of the cycle space of K_5 is 6, B consists of 6 cycles $Z_1, Z_2, \ldots, Z_6$. Since each Z_i contains 3 or more edges, the number of edges in Z_i totals at least 18. On the other hand K_5 has 10 edges, each of which is contained in at most two cycles of B. Therefore at least 8 edges of K_5 appear twice in these cycles. Then the sum $Z_1 \oplus Z_2 \oplus \cdots \oplus Z_6$ consists of at most two edges. Since $Z_1, \ldots, Z_6$ are independent, the sum must be nonzero and hence consist of at least three edges, a contradiction. Similarly one can show that $K_{3,3}$ has no 2-basis.   Q.E.D.

Lemma 1.6. *No subdivision of K_5 or $K_{3,3}$ has a 2-basis.*

Proof. If a subdivison of K_5 or $K_{3,3}$ contained a 2-basis B, then the collection of cycles in K_5 or $K_{3,3}$ which naturally correspond to the cycles in B would be a 2-basis of K_5 or $K_{3,3}$, contradicting Lemma 1.5.

Lemma 1.7. *Let G be a graph, and let e be an edge of G. If G has a 2-basis B then $G' = G - e$ also has a 2-basis.*

Proof. Note that the collection of cycles in G' consists of all the cycles in G not containing e. Since B is a 2-basis of G, e is contained in one or two cycles of B. Thus there are two cases to consider.

Case 1: e is contained in only one cycle Z of B. Clearly the cycles in $B - \{Z\}$ are independent since the cycles in B are so. Since B is a cycle basis of G, any cycle C of G is a sum of cycles in B. Therefore, if C does not contain e, that is, C is a cycle of G', then C is a sum of cycles in $B - \{Z\}$. Thus $B - \{Z\}$ is a cycle basis of G', and hence clearly it is a 2-basis of G'.

Case 2: e is contained in two cycles Z and Z' of B. In this case the sum $Z \oplus Z'$ is decomposed into edge-disjoint cycles $Z_1, Z_2, \ldots, Z_r$. Similarly as in Case 1 one can easily show that $B' = (B - \{Z, Z'\}) \cup \{Z_1, \ldots, Z_r\}$ is a 2-basis of G'.
                                                                          Q.E.D.

Lemmas 1.6 and 1.7 above imply the following MacLane's criterion.

Theorem 1.5. *A 2-connected graph G is planar if and only if G has a 2-basis.*

Proof. Since we have already seen the necessity, we shall establish the sufficiency. Suppose that a 2-connected graph G has a 2-basis, but is nonplanar. Then Kuratowski's theorem (Theorem 1.3) implies that G has a subdivision G' of K_5 or $K_{3,3}$. Repeated application of Lemma 1.7 implies that G' has a 2-basis. However by Lemma 1.6 G' has no 2-basis, a contradiction.   Q.E.D.

We are now ready to prove the sufficiency of Theorem 1.4 (Whitney's criterion): if a 2-connected graph G has a combinatorial dual G^* then G is planar.

Proof of sufficiency of Theorem 1.4. Suppose that a 2-connected graph G has a combinatorial dual $G^* = (V^*, E^*)$. Since G is 2-connected, every two edges of G lie on a common cycle, and hence every two edges of G^* lie on a common cutset. Therefore G^* must be 2-connected. Let S_v, $v \in V^*$, be the set of edges of G^* incident to vertex v, then S_v is a cutset of G^* since G^* is 2-connected. Any cutset S of a graph G^* is a sum of S_v over all vertices v in a component of $G^* - S$. Thus the cutset space of G^* is generated by all these S_v. Let u be an arbitrary vertex of G^*, then S_u is a sum of S_v over all v in $V^* - u$. Thus $B^* = \{S_v : v \in V^* - u\}$ is a basis of the cutset space of G^*. Obviously every edge of G^* is contained in at most two S_v of B^*. Thus the collection of cycles in G corresponding to cutsets of B^* is a 2-basis of G. Therefore by Theorem 1.5 G is planar. Q.E.D.

1.7. Bounds for planar graphs

In Section 1.4 we showed that nontrivial bounds on the number m of edges in a planar graph follow from Euler's formula. In this section we present several interesting bounds on other graph invariants for planar graphs, which will be used by algorithms in later chapters.

An *independent set* of a graph G is a set of nonadjacent vertices, and a *maximum independent set* is one of maximum cardinality. For example the graph in Fig. 1.1(b) has a maximum independent set $\{v_5, v_6\}$. The problem of finding a maximum independent set is "*NP*-hard" even for planar graphs, and no algorithm is known for finding a maximum independent set in polynomial-time. There may exist a large independent set in a planar graph if the minimum degree is 1 or 2. For example the complete bipartite graph $K_{2,n-2}$, $n \geqslant 4$, which is a planar graph with minimum degree 2, contains a maximum independent set of $n - 2$ vertices. This is not the case if the minimum degree is 3 or more. Corollary 1.2 together with an easy observation provide nontrivial upper bounds on the number of vertices in an independent set for such planar graphs. Note that the minimum degree of a planar graph is necessarily 5 or less as shown in Corollary 1.4.

Theorem 1.6. *If G is a planar connected graph with minimum degree 3, 4, or 5, then a maximum independent set of G contains less than $2n/3$, $2n/4$, or $2n/5$ vertices, respectively.*

Proof. Let $I^* \subset V$ be a maximum independent set of G. Let G' be the spanning subgraph obtained from G by deleting all the edges having both ends in $V - I^*$, and assume that G' has m' edges. Since G' is a planar bipartite graph with n (≥ 4) vertices, Corollary 1.2 implies $m' \leq 2n - 4$. On the other hand, since $d(G, v) = d(G', v)$ for every vertex v in I^*, we have $m' \geq 3|I^*|$, $4|I^*|$, or $5|I^*|$ according to the minimum degree. The two inequalities above immediately imply the claim.   Q.E.D.

A *matching* of a graph G is a set of nonadjacent edges, and a *maximum matching*, denoted by $M(G)$, is one of maximum cardinality. For example for G in Fig. 1.1(b) both $\{(v_1, v_2), (v_3, v_4)\}$ and $\{(v_1, v_5), (v_2, v_3), (v_4, v_6)\}$ are matchings, and the latter is a maximum. Unlike the maximum independent set, efficient polynomial-time algorithms have been obtained for finding a maximum matching in general graphs [EK75, MV80]. Similarly as in Theorem 1.6 we can obtain a lower bound on $|M(G)|$ for planar graphs with minimum degree 3 or more. To prove the bound we need the following sophisticated characterization on $|M(G)|$, due to Tutte [Tut47] and Berge [Ber57], where $q(G)$ denotes the number of *odd components* of G, that is, ones having an odd number of vertices.

Theorem 1.7. *Let G be a graph of n vertices. If*

$$u = \max_{S \subset V} \{q(G - S) - |S|\}, \tag{1.1}$$

then $|M(G)| = (n - u)/2$, that is, there are u vertices with which no edge of $M(G)$ is incident.

Proof. See [Bol79 or Beg76].

Theorem 1.7 together with Corollary 1.2 yield the following lower bound on $|M(G)|$ for planar graphs [Nis79, NB79]. Throughout this book $\lfloor x \rfloor$ denotes the largest integer $\leq x$, while $\lceil x \rceil$ is the smallest integer $\geq x$.

Theorem 1.8. *If a planar connected graph G has minimum degree 3 or more, then*

$$|M(G)| \geq \min\{\lfloor n/2 \rfloor, \lceil (n + 2)/3 \rceil\}.$$

Proof. By Theorem 1.7 it suffices to show that if $u \geq 2$ then $u \leq \lfloor (n - 4)/3 \rfloor$. Suppose that $u \geq 2$, and let S be the subset of V attaining the maximum in (1.1), then clearly $|S| \geq 1$. Assume that $G - S$ has q_1 odd components containing exactly one vertex. There are two cases to consider.

Case 1: $q_1 = 0$. In this case $n \geqslant 3q(G - S) + |S|$ since every odd component contains three or more vertices. Then the desired inequality follows immediately:

$$u = q(G - S) - |S| \leqslant (n - |S|)/3 - |S| \leqslant (n - 4)/3.$$

Case 2: $q_1 \geqslant 1$. Remove from G all the vertices in the component of $G - S$ having two or more vertices, and delete all the edges with two ends in S. Let G' be the resulting graph, and let us denote by n' and m' the numbers of vertices and edges in G', respectively. Clearly

$$n' = q_1 + |S|, \tag{1.2}$$

and

$$n' \leqslant n - 3(q(G - S) - q_1),$$

which implies

$$q(G - S) - q_1 \leqslant (n - n')/3. \tag{1.3}$$

Since G' is a bipartite planar graph with n' ($\geqslant 4$) vertices and all the q_1 vertices in one partite set have degree 3 or more, Corollary 1.2 and an easy counting yield

$$3q_1 \leqslant m' \leqslant 2n' - 4. \tag{1.4}$$

We have from (1.2) and (1.4)

$$6q_1 \leqslant 4n' - 8 = n' - 8 + 3(q_1 + |S|),$$

which implies

$$q_1 - |S| \leqslant (n' - 8)/3. \tag{1.5}$$

The desired inequality immediately follows from (1.3) and (1.5):

$$u = (q_1 - |S|) + (q(G - S) - q_1) \leqslant (n - 8)/3.$$

Q.E.D.

A *vertex cover* of a graph G is a set of vertices such that every edge of G is incident on at least one of the vertices. A *minimum vertex cover* is one of minimum cardinality. For example $\{v_1, v_2, v_3, v_4\}$ is a minimum vertex cover of G in Fig. 1.1(b). Every vertex cover must contain either of the two ends of each edge in a matching. Therefore the cardinality of the minimum vertex cover of G is no less than $|M(G)|$. Thus the following corollary is an immediate consequence of Theorem 1.8.

Corollary 1.5. *If a planar connected graph G has minimum degree 3 or more, then the cardinality of a minimum vertex cover is at least* $\min\{\lfloor n/2 \rfloor, \lceil (n + 2)/3 \rceil\}$.

CHAPTER 2

ALGORITHMIC FOUNDATIONS

2.1. What is an algorithm?

Consider a computational problem on graphs, such as the *planarity testing problem*: given a graph, is it planar? If the given graph is small, one may perform the test by hand, as we did in Section 1.1 for the graph depicted in Fig. 1.1(a). However the problem would not be solved without the existence of fast digital computers if the graph has hundreds or thousands of vertices. Thus we need an "*algorithm*": a precise method usable by a computer to solve a problem. An algorithm must solve any instances of a problem and terminate after a finite number of operations.

Direct application of Kuratowski's theorem (Theorem 1.3) provides the following procedure for the planarity testing problem:

Systematically generate all (edge-induced) subgraphs H of a given graph G, and check whether each of the H's is a subdivision of K_5 or $K_{3,3}$. If there is at least one such H, G is nonplanar; otherwise G is planar.

Clearly a graph G with m edges has 2^m subgraphs, and it is easy to check whether each of them is a subdivision of K_5 or $K_{3,3}$. Thus the procedure above can test the planarity for *any* given graph and terminate within finite time. That is, it is indeed an algorithm, although it is not efficient.

Let us consider the following classical question: are there computational problems for which there is no algorithm? A. T. Turing [Tur36] introduced a mathematical model of computers, called a *Turing machine*, and showed that such unsolvable problems do exist. A typical example is the so-called halting problem: given a computer program with its input, will it ever halt? Turing proved that there is *no* algorithm that solves correctly all instances of this problem within finite time. Thus there exist problems for which there is no algorithm. However all of the problems discussed in this book *have* algorithms.

2.2. Machine model and complexity

The planarity testing algorithm mentioned in Section 2.1 would require at least 2^m steps (elementary instructions). Therefore the solution by this algorithm of a modestly sized graph, say having 100 edges, would require more than one hundred centuries, even under the most optimistic assumptions about the speed of computers in the future (2^{100} has about 30 decimal digits). Thus that algorithm is completely useless in practice. In contrast efficient algorithms for the problem are known as we will see in Chapter 3.

In order to study the efficiency of algorithms, we need a model of computation. The earliest and simplest one is the Turing machine. The model was very useful in high-level theoretical studies of computation, such as the existence of unsolvable problems. However the model is not realistic enough to allow accurate analysis of practical algorithms. In this book we use the so-called *random-access machine* (RAM) model, introduced by Cook and Reckhow [CR76]. The RAM is an abstraction of a general-purpose digital computer in which each memory cell has a unique address and can store a single integer or real number, and it sequentially performs an access to any cell, an arithmetic operation, or a Boolean operation, following a finite sequence of instructions, called a program. We assume that each of the operations requires *unit time*, and each of the memory cells uses *unit space*. That is, our model is the *unit-cost* RAM. This assumption is realistic for all algorithms discussed in this book.

The most widely accepted complexity measures for an algorithm are the running time and storage space. The running time is the number of operations it performs before producing the final answer, while the storage space is the number of memory cells it uses. The number of operations or cells required by an algorithm is not the same for all problem instances. For example, in the planarity testing, the number of operations required to test the planarity of a graph with n vertices and m edges may vary considerably with graphs, even if n and m are kept constant. Thus we consider all inputs of a given size together, and we define the complexity of the algorithm for that input size to be the worst-case behavior of the algorithm on any of these inputs. Then the running time (or storage space) is a function of the size n of the input, such as $100n$, $n \log n$, $10n^{3.5}$, and 2^n.

In analyzing the complexity of an algorithm, we are often interested only in the "asymptotic behavior", that is, the behavior of the algorithm when applied to very large inputs. Fig. 2.1 illustrates a table of running times for algorithms of various complexity. As can be seen from the table, the running time $1000n$ is much smaller than that of n^3. Thus we are interested in the "order" of the complexity of algorithms. To deal with such a property of functions we shall use the following notation for asymptotic running time or space. Let $f(n)$ and $g(n)$ be functions from the positive integers to the positive reals, then we write $f(n) = O(g(n))$ if there exists positive constants c_1 and c_2 such that $f(n) \leqslant c_1 g(n) + c_2$ for all n. Thus the running time of an algorithm may be bounded from above by phrasing like "take time $O(n^2)$".

Time	Size			
	10	100	500	1000
$1000n$	0.01 s	0.1 s	0.5 s	1 s
$100n \log n$	0.003 s	0.06 s	0.45 s	1 s
n^3	0.001 s	1 s	2 min	16 min
2^n	0.001 s	40 000 CENT		

Fig. 2.1. Running time estimate. (One step takes one microsecond.)

An algorithm is said to be *polynomially bounded* (or simply *polynomial*) if its complexity is bounded above by a polynomial of the size n of a problem instance. Examples of such complexity are n, $n \log n$, $n^{1.5}$, and n^{100}. The remaining algorithms are usually referred to as *exponential* (or *nonpolynomial*). Examples of such complexity are 2^n, $n!$, and n^n. As the problem size increases, polynomial algorithms become gradually unusable, whereas exponential algorithms rapidly become completely useless, due to the so-called combinatorial explosion. Thus there is a big discrepancy between the two classes of algorithms, and it is a general agreement that an algorithm is practically useful only if it is polynomially bounded.

2.3. NP-complete

There are a number of interesting computational problems for which it has not been proved whether there is a polynomial algorithm or not. Most of them are "*NP-complete*", which we will briefly explain in this section.

The state of algorithms consists of the current values of all the variables and the location of the current instruction to be executed. A *deterministic algorithm* is one for which each state, upon execution of the instruction, uniquely determines at most one following state (next state). All computers, which exist now, run deterministically. A problem Q is in the *class P* if there exists a deterministic polynomial-time algorithm which solves Q.

In contrast a *nondeterministic algorithm* is one for which a state may determine many next states simultaneously. We may regard a nondeterministic algorithm as having the capability of branching off into many copies of itself, one for each next state. Thus, while a deterministic algorithm must explore a set of alternatives one at a time, a nondeterministic algorithm examines all alternatives at the same time. A problem Q is in the *class NP* if there exists a nondeterministic polynomial-time algorithm which solves Q. Clearly $P \subset NP$.

An example of a problem in NP is the *travelling salesman problem*: given n cities and pairwise distances between them, is there a tour that passes through each city once, returns to the starting point, and has a total length of at most a given value x? Select nondeterministically each of the $n!$ possible tours, and check whether it has a length of at most x; if so, answer YES. This nondeterministic algorithm solves the travelling salesman problem in polynomial time, so the problem is in NP.

Another example of problems in NP is the *nonplanarity testing problem*: given a graph, is it nonplanar? Clearly the algorithm based on Kuratowski's theorem gives a nondeterministic polynomial algorithm for the problem. However, as we will see in Chapter 3, the planarity testing problem and hence the nonplanarity testing problem can be solved in linear time by a deterministic algorithm. Thus both are indeed in P.

Among the problems in NP are those that are hardest in the sense that if one can be solved in polynomial-time then so can every problem in NP. These are called the *NP-complete problems*. Cook [Coo71] formalized this notion and illustrated it with several NP-complete problems. Karp [Kar72] showed that a number of important problems, including the Hamiltonian cycle problem, are NP-complete. At present hundreds of problems are known to be NP-complete [GJ79]. Examples of known NP-complete problems are the travelling salesman problems, integral linear programming, the satisfiability problem, the independent vertex set problem, and the graph coloring problem. This class of NP-complete problems has the following very interesting properties.

(a) No NP-complete problem can be solved by any known polynomial algorithm (although many brilliant researchers have struggled to solve them for many years).

(b) If there is a polynomial algorithm for any NP-complete problem, then there are polynomial algorithms for all NP-complete problems.

As a result of these two facts, many people have conjectured that there can be no polynomial algorithm for any NP-complete problem. However nobody has been able to prove this. Thus the most important open problem in theoretical computer science is to determine whether $P = NP$, although the solution seems to need the development of entirely new mathematical techniques.

Sometimes we may be able to show that, if problem Q is solvable in polynomial time, all problems in *NP* are so, but we are unable to argue that $Q \in NP$. So Q does not qualify to be called *NP*-complete. Yet, undoubtedly Q is as hard as any problem in *NP*. Such a problem Q is called *NP-hard*. One example is the minimum travelling salesman problem, in which one would like to find a tour of minimum length. It has been established that a number of practical optimization problems are *NP*-hard [GJ79].

2.4. Data structure and graph representation

In this section we first give a brief account of basic data structures, and then show two methods to represent a graph in a computer. A vector or a set of variables is usually stored as a (*1-dimensional*) *array*, and a matrix as a *2-dimensional array*. The main feature of an array is its index capability. The indices should uniquely determine the location of each entry, and accessing an entry is done in a constant amount of time.

A *list* is a data structure which consists of homogeneous records which are linked together in a linear fashion. Each record contains one or more items of data and one or more of pointers. Fig. 2.2(c) shows five singly linked lists; each record has a single forward pointer indicating the address of the memory cell of the next record. In a *doubly linked list* each record has forward and backward pointers, and we can delete a record or insert a record without being given the location of the previous record.

A stack and a queue are two special types of lists. A *stack* is a list in which we are only permitted to insert and delete elements at one end, called the *top* of the stack. Thus a stack functions in a last-in, first-out manner with respect to insertion and deletion. A *queue* is a list in which we are only permitted to insert at one end, called the *tail* of the queue, and delete from the other end, called the *head* of the queue. That is, a queue functions in a first-in, first-out manner. Both stack and queue are realized by a list with pointers indicating the current locations of the top, head or tail.

We measure the complexity of an algorithm as a function of the size of the input of an algorithm. But what is the size of the input of a graph problem? To represent a graph by a computer, we must encode it as a sequence of symbols over some fixed alphabet such as bits or typewriter symbols. The size is the length of the sequence.

A graph may be represented in many ways. For example we can associate with a graph $G = (V, E)$ its $n \times n$ *adjacency matrix* $A = [a_{ij}]$ such that $a_{ij} = 1$ if $(v_i, v_j) \in E$, and $a_{ij} = 0$ otherwise. Fig. 2.2 illustrates a graph (a) and its adjacency matrix (b). By definition the main diagonal of A is all zeros, and A is symmetric. An adjacency matrix uses n^2 space to represent any graph of n

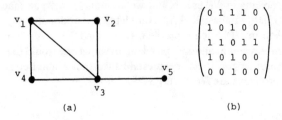

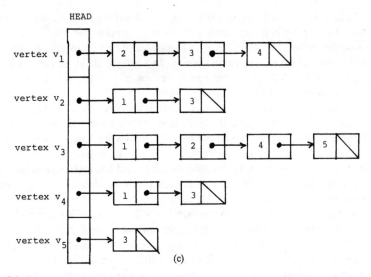

Fig. 2.2. Representation of a graph: (a) graph, (b) adjacency matrix, and (c) adjacency lists.

vertices. It is not economical when a graph is *sparse*, that is, the number m of edges is far less than $n(n-1)/2$. In this case simply listing the edges one by one would be much more efficient. If A is stored in a computer as a 2-dimensional array, then only one step is required for the statement "Is $(v_i, v_j) \in E$?" or "Erase the edge (v_i, v_j)". However, scanning the neighbours of a vertex v requires n steps even if degree $d(v)$ is far less than n.

Another useful way of representing a graph is by its *adjacency lists*. For each vertex $v \in V$, we record the set $N(v)$ of v's neighbours. The sets are stored in a computer as lists Adj(v) (see Fig. 2.2(c)). The space requirement for the adjacency lists is

$$O\left(\sum_{v \in V} [1 + d(v)] \right) = O(n + m).$$

Thus the representation is much more economical than the adjacency matrix if a graph is sparse. Scanning the adjacency list Adj(v) can be done in $d(v)$ steps, but the statement "Is $(v_i, v_j) \in E$?" requires $d(v)$ steps.

We can assume that $m \geqslant n/2$ (for instance, if our graph has no isolated vertices). Therefore m is a reasonable approximation to the size of a graph. For a planar graph we usually use n as the approximated size of an input graph since $m \leqslant 3n$ by Corollary 1.1. Thus we analyse the complexity of algorithms using n (for planar graphs) or n and m (for general graphs) as parameters.

A graph algorithm is said to be *linear* if it runs in $O(n + m)$ times on a graph with n vertices and m edges. This is usually the best that one could expect for a graph algorithm.

2.5. Exploring a graph

When designing algorithms on graphs, we often need a method for exploring the vertices and edges of a graph. Most important are the depth-first search (DFS) and breadth-first search (BFS). In both methods each edge is traversed exactly once in the forward and reverse directions and each vertex is visited. Thus both run in linear time. The choice of which method to use must depend on the problem. DFS has been used for planarity testing, certain connectivity problems (partitioning a graph into blocks or 3-connected components), etc. BFS has been used for shortest path problems, network flow problems, etc.

2.5.1. Depth-first search

Consider visiting the vertices of a graph in the following way. We select and visit a starting vertex v. Then we select any edge (v, w) incident on v and visit w. In general, suppose x is the most recently visited vertex. The search is continued by selecting some unexplored edge (x, y) incident on x. If y has been previously visited, we find another new edge incident on x. If y has not been previously visited, then we visit y and begin the search anew starting at y. After completing the search through all paths beginning at y, the search returns to x, the vertex from which y was first reached. The process of selecting unexplored edges incident to x is continued until the list of these edges is exhausted. This method is called the *depth-first search* since we continue searching in the deeper direction as long as possible.

A depth-first search of an undirected graph $G = (V, E)$ partitions E into two sets T and B, where T comprises a spanning forest of G. The edge (x, y) is placed into T if vertex y was visited for the first time immediately following a visit to x. The edges in T are called *tree edges*. The remaining edges, called *back edges*, are placed into B. The algorithm is given below.

procedure DEPTH-FIRST-SEARCH:
begin
   $T := \varnothing$;
   for all v in V *do* mark v "new";
   while there exists a vertex v in V marked "new" *do*
     SEARCH(v)
end.

procedure SEARCH(v):
begin
  mark v "old";
  for each vertex w in Adj(v) *do*
   if w is marked "new" *then*
    begin
      add (v, w) to T;
      SEARCH(v)
    end
end;

For example, the depth-first search starting with vertex v_1 of graph $G = (V, E)$ in Fig. 2.2(a) partitions E into the spanning tree T and the set B of back edges; T and B are drawn by solid and dotted lines in Fig. 2.3, respectively.

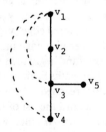

Fig. 2.3. DFS tree T.

2.5.2. Breadth-first search

In implementing the breadth-first search (BFS) we choose an arbitrary vertex and put it on a queue of vertices to be visited. We repeatedly delete the vertex x at the head of the queue Q, and scan the adjacency list Adj(x) with inserting to Q all neighbours of x which have never been inserted to Q. For simplicity we may assume that a given graph $G = (V, E)$ is connected; otherwise repeatedly apply BFS to each component. BFS partitions the edges E of a connected graph into the spanning tree T and the set B of back edges. The algorithm is given below.

procedure BREADTH-FIRST-SEARCH
begin
   $T := \varnothing$; $Q := $ empty queue;
   for all $v \in V$ *do* mark v "not reached"
   choose an arbitrary vertex $r \in V$ as a starting vertex and add r to queues Q;
   $\{r$ is a root of tree $T\}$
   mark r "reached";
   LEVEL$(r) := 0$;
   while Q is nonempty
     begin
      $x :=$ head of Q; $Q := Q - x$;
      SCAN(x)
     end
end.

procedure SCAN(x):
begin
  for each $y \in$ Adj(x) *do*
   if y is marked "not reached" *then*
    begin
     add the edge (x, y) to T;
     LEVEL$(y) :=$ LEVEL$(x) + 1$;
     add y to queue Q
     mark y "reached"
    end
end.

Thus the search scans the adjacency lists of vertices in a first reached, first scanned manner; in this sense it is called the breadth-first search. The breadth-first search tree T and the vertex levels satisfy the following properties:

(a) Every edge of G, whether tree or back edge, joins two vertices whose levels
    differ by at most one;
(b) The level of v equals the length (i.e. number of edges) of the shortest path
    from the root r to v in G.

As well as DFS, BFS explores each edge exactly twice, in the forward and
backward directions, and hence runs in linear time.

The breadth-first search starting with vertex v_1 of a graph $G = (V, E)$ in
Fig. 2.2(a) partitions E into the spanning tree T and the set B of back edges. T
and B are drawn by solid and dashed lines in Fig. 2.4, where the numbers next
to vertices are levels.

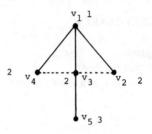

Fig. 2.4. BFS tree T.

CHAPTER 3

PLANARITY TESTING AND EMBEDDING

3.1. Introduction

There are many practical situations in which one may wish to determine whether a given graph is planar, and if so, to find a planar embedding (drawing) of the graph. For example, in the layout of printed or VLSI circuits, one is interested in knowing whether a graph G representing a circuit is planar and also in finding a planar embedding of G if it is planar. In this chapter, we present linear algorithms for the problem, due to [LEC67, BL76, CNAO85].

Two planarity testing algorithms which run in linear time have been established: one by Hopcroft and Tarjan [HT74], and the other by Booth and Lueker [BL76]. We present the latter called the "vertex addition algorithm" in Section 3.2 since it is conceptually simpler than the former.

The vertex addition algorithm was first presented by Lempel, Even and Cederbaum [LEC67], and improved later into a linear algorithm by Booth and Lueker [BL76] employing an "st-numbering" algorithm and a data structure called a "PQ-tree." The algorithm adds one vertex in each step. Previously embedded edges incident on this vertex are connected to it, and new edges incident on it are embedded and their ends are left unconnected. Sometimes whole pieces have to be reversed (flipped) around or permuted so that some ends occupy consecutive positions. If the representation of the embedded subgraph is updated with each alteration of the embedding, then the final representation will be an actual embedding of a given whole graph. Thus it is not difficult to derive an $O(n^2)$ time embedding algorithm from the testing algorithm.

In Section 3.3 we present a simple linear algorithm for embedding planar graphs, which is based on the vertex addition algorithm. The embedding algorithm, due to Chiba et al. [CNAO85], runs in two phases. In the first phase the algorithm embeds a digraph obtained from a given planar (undirected) graph by assigning a direction to every edge from the end having a greater *st*-number to the other end. In the second phase the algorithm extends the embedding of the digraph into an embedding of the given planar (undirected) graph.

3.2. Planarity testing

We first define some terms and then present the vertex addition algorithm for testing planarity. Let $G = (V, E)$ be a graph with vertex set V and edge set E. A graph G is represented by a set of n lists, called "adjacency lists"; the list Adj(v) for vertex $v \in V$ contains all the neighbours of v. For each $v \in V$ an actual drawing of a planar graph G determines, within a cyclic permutation, the order of v's neighbours embedded around v. *Embedding* a planar graph G means constructing adjacency lists of G such that, in each Adj(v), all the neighbours of v appear in clockwise order with respect to an actual drawing. Such a set Adj of adjacency lists is called an *embedding* of G. One example is illustrated in Fig. 3.1(d) which is an embedding of a graph G in Fig. 3.1(a).

Planarity testing determines whether a given graph is planar or not, that is, whether a given graph has an embedding on the plane. A graph G is planar if and only if the 2-connected components of G are planar [Har72]. Moreover, one can easily obtain an embedding of the entire graph G from embeddings of all the 2-connected components of G. Hence we assume that the given graph G is 2-connected. Due to Corollary 1.1 we may further assume that G satisfies $m \leqslant 3n$; otherwise G is nonplanar.

The "*st*-numbering" plays a crucial role in the testing algorithm. A numbering of the vertices of G by $1, 2, \ldots, n$ is called an *st-numbering* if the two vertices "1" and "n" are necessarily adjacent and each j of the other vertices is adjacent to two vertices i and k such that $i < j < k$. The vertex "1" is called a *source* and is denoted by s, while the vertex "n" is called a *sink* and is denoted by t. Fig. 3.1(a) illustrates an *st*-numbering of a graph. Every 2-connected graph G has an *st*-numbering, and an algorithm given by Even and Tarjan [ET76] finds an *st*-numbering in linear time as described below.

We first present a procedure DFS which executes the depth-first search with choosing an arbitrary edge (t, s) as the first edge, and computes, for each vertex v, its depth-first number DFN(v), its father FATH(v) and its lowpoint LOW(v). These values are used in the *st*-numbering algorithm. The lowpoint is defined as follows:

LOW(v) = min({v} $\cup$ {w | there exists a back edge (u, w) such that u is a descendant of v and w is an ancestor of v in a DFS tree T}),

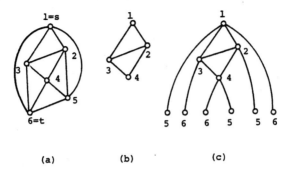

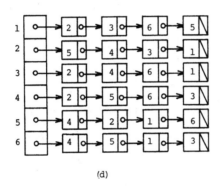

(d)

Fig. 3.1. Illustrations of: (a) *st*-numbered graph G, (b) G_4, (c) bush form B_4 and (d) adjacency lists of embedded G.

where we assume that the vertices v are named by their depth-first numbers DFN(v). This definition is equivalent to

$$LOW(v) = \min(\{v\} \cup \{LOW(x) \mid x \text{ is a son of } v\}$$
$$\cup \{w \mid (v, w) \text{ is a back edge}\}).$$

Fig. 3.2 illustrates these values. The following depth-first search algorithm DFS, a variant of one in Chapter 2, computes the required values.

procedure DFS(G);
  procedure SEARCH(v);
    begin
      mark v "old";

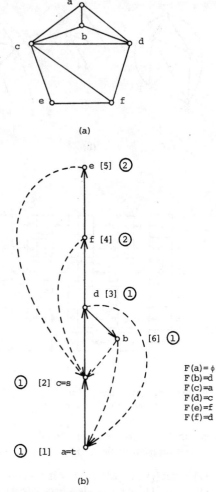

Fig. 3.2. (a) Graph G and (b) DFS-tree T, and DFN, LOW and FATH;
tree edges are drawn by solid line, and back edges by dotted lines;
DFNs are written in brackets, while LOW numbers are in circles.

DFN(v) := COUNT;
COUNT := COUNT + 1;
LOW(v) := DFN(v);
for each vertex w on Adj(v)

```
        do if w is marked "new"
          then begin
            add (v, w) to T;
            FATH(w) := v;
            SEARCH(w);
            LOW(v) := min{LOW(v), LOW(w)}
            end
          else if w is not FATH(v)
            then LOW(v) := min{LOW(v), DFN(w)}
      end;
begin
    T := ∅; {T is a DFS-tree}
    COUNT := 1;
    mark each vertex of G "new";
    let v be any vertex of G;
    SEARCH(v)
end.
```

We next present a function PATH. Initially, the two vertices s and t together with the first searched edge (s, t) are marked "old" and all the other vertices and edges are marked "new". Note that $DFS(s) = 2$ and $DFS(t) = 1$. PATH(v) takes as the value a path going from v to an "old" vertex. The procedure PATH(v) is as follows:

Case 1: there is a "new" back edge (v, w).
  mark (v, w) "old";
  PATH := vw;
  return.
Case 2: there is a "new" tree edge (v, w).
  let $w_0(= w)w_1w_2\cdots w_k(= \text{LOW}(w))$ be the path which defined LOW(w), that is, it runs up the tree and ends with a back edge into a vertex u such that $DFN(u) = LOW(w)$;
  PATH := $vw_0w_1\cdots w_k$;
  mark all the vertices and edges on the path "old";
  return.
Case 3: there is a "new" back edge (w, v).
  {in this case $DFN(w) > DFN(v)$};
  let $w_0(= w)w_1\cdots w_k$ be the path going backward on tree edges to an "old" vertex; {we can trace the path using father pointers FATH};
  PATH := $vw_0w_1\cdots w_k$;
  mark all the vertices and edges on the path "old";

return.
Case 4: otherwise, that is, all edges (v, w) incident to v are "old".
PATH $:= \varnothing$;
return.

We now give one example. Mark vertices a and c and an edge (a, c) of G in Fig. 3.2 "old" and repeat to apply the function PATH(c) to G until PATH(c) $= \varnothing$, then we successively get paths

$$\text{PATH}(c) = cda, \tag{1}$$

$$\text{PATH}(c) = cbd, \tag{2}$$

$$\text{PATH}(c) = cefd, \quad \text{and} \tag{3}$$

$$\text{PATH}(c) = cf \tag{4}$$

(see Fig. 3.3). Then, applying PATH(b), we have

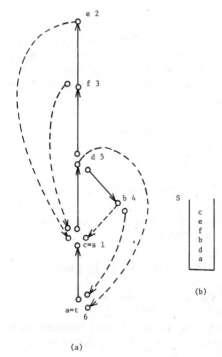

(a)

Fig. 3.3. Illustrations of: (a) paths in G in Fig. 3.1 and st-numbers, and (b) stack S.

$$\text{PATH}(b) = ba. \tag{5}$$

The path (1) applies to Case 2, the paths (2), (3) and (4) to Case 3 and the path (5) to Case 1. Thus such iterative applications of PATH partition the edges of a graph into several paths.

We are now ready to present the st-numbering algorithm.

procedure ST-NUMBER(G);
  begin
    mark s, t and (s, t) "old" and all the other vertices and edges "new";
    push down t and s into stack S in this order;
      $\{s$ is over $t\}$
    COUNT := 1;
    pop up the top entry v from S;
    while $v \neq t$
      do begin
        if PATH(v) = $\varnothing$
          then begin
            STN(v) := COUNT; COUNT := COUNT + 1
          end
          else begin
            let PATH(v) = $vu_1u_2 \cdots u_kw$;
            push down the vertices $u_k, u_{k-1}, \ldots, u_1, v$ into S in this order
            $\{v$ is a top entry of $S\}$
          end;
        pop up the top entry v from S
      end;
    STN(t) := COUNT
  end.

Assume that we apply ST-NUMBER to G in Fig. 3.2 letting $s = c$ and $t = a$ and that iterative calls for PATH in the algorithm produce a sequence of paths (1)–(5) above. Then the stack S will become as illustrated in Fig. 3.3(b) just after the path (3) is processed, and st-numbers will be assigned as in Fig. 3.3(a).

We have the following theorem on the algorithm.

Theorem 3.1. *The algorithm ST-NUMBER computes correctly an st-numbering for a 2-connected graph G.*

Proof. Every vertex v is numbered and permanently removed from S only after all the edges incident to v become "old." Since G is 2-connected, every vertex v is reachable from s by a path not passing through t. Therefore every vertex is placed on S before t is removed, and hence every vertex is numbered. Thus we shall show that the numbering is indeed an st-numbering. Clearly $STN(s) = 1$ and $STN(t) = n$. Every other vertex v is placed on S, for the first time, as an intermediate vertex on a path. Moreover, the last vertex w of the path marked "old" has already been pushed and yet stays in S, since w has been incident to a "new" edge. Thus, two neighbours of v are stored in S below and above v, and hence one above v is assigned a lower number and one below v a higher number. Thus the numbering is indeed an st-numbering.   Q.E.D.

From now on we refer to the vertices of G by their st-numbers. Let $G_k = (V_k, E_k)$ be the subgraph of G induced by the vertices $V_k = \{1, 2, \ldots, k\}$. If $k < n$ then there must exist an edge of G with one end in V_k and the other in $V - V_k$. Let G'_k be the graph formed by adding to G_k all these edges, in which the ends in $V - V_k$ of added edges are kept separate. These edges are called *virtual edges*, and their ends in $V - V_k$ are called *virtual vertices* and labelled as their counterparts in G, but they are kept separate. Thus there may be several virtual vertices with the same label, each with exactly one entering edge. Let B_k be an embedding of G'_k such that all the virtual vertices are placed on the outer face. B_k is called a *bush form* of G'_k. The virtual vertices are usually placed on a horizontal line. G, G_k, and B_k are illustrated in Fig. 3.1. The following lemma implies that every planar graph G has a bush form B_k for $1 \le k \le n$.

Lemma 3.1. [*Eve* 79] *Let* $1 \le k \le n$. *If edge* (s, t) *is drawn on the boundary of the outer face in an embedding of* G, *then all the vertices and edges of* $G - G_k$ *are drawn in the outer face of the plane subgraph* G_k *of* G.

Proof. Assume that some vertices of $G - G_k$ lie on a face F of G_k. Since all these vertices are higher than the vertices on F, the highest one must be a sink. Thus the face F must be the outer face of the plane graph G_k.   Q.E.D.

An *upward digraph* D_u is defined to be a digraph obtained from G by assigning a direction to every edge so that it goes from the larger end to the smaller. An *upward embedding* A_u of G is an embedding of the digraph D_u. In an embedding of an *undirected* graph G, a vertex "v" appears in list $\text{Adj}(w)$ and "w" appears in list $\text{Adj}(v)$ for every edge (v, w). However, in an upward embedding A_u of G, the head "w" appears in adjacency list $A_u(v)$ but the tail "v" does not appear in $A_u(w)$ for every directed edge $\langle v, w \rangle$. Fig. 3.4 depicts an upward digraph D_u and an upward embedding A_u for the graph G in Fig. 3.1(a). We use a special data structure "*PQ-tree*" to represent B_k. A *PQ-tree*

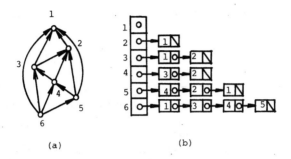

Fig. 3.4. Illustrations of: (a) upward digraph D_u, and (b) upward embedding A_u for a graph G in Fig. 3.1(a).

consists of "P-nodes", "Q-nodes" and "leaves". A P-node represents a cutvertex of B_k, so the sons of a P-node can be permuted arbitrarily. A Q-node represents a 2-connected component of G_k, and the sons of a Q-node are allowed only to reverse (flip over). A leaf indicates a virtual vertex of B_k. In an illustration of a PQ-tree, a P-node is drawn by a circle and a Q-node by a rectangle. A bush form B_k and a PQ-tree representing B_k are illustrated in Fig. 3.5. Thus a PQ-tree represents all the permutations and the reversions possible in a bush form B_k.

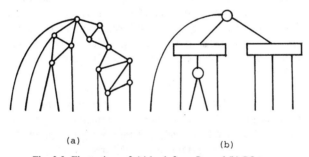

Fig. 3.5. Illustrations of: (a) bush form B_k, and (b) PQ-tree.

The key idea of the vertex addition algorithm is to reduce the planarity testing of G_{k+1} to the problem which asks for permutations and reversions to make all the virtual vertices labelled "$k + 1$" occupy consecutive positions. The following lemma guarantees that this reduction is possible.

Lemma 3.2. [Eve 79] *Let B_k be any bush form of subgraph G_k of a planar graph G. Then there exists a sequence of permutations and reversions to make all the* *virtual vertices labelled "$k + 1$" occupy consecutive positions on the horizontal line.*

Proof. We prove a more general proposition: if B and B' are two sub-bush forms, having the same vertices, of any distinct bush forms of G_k, then there exists a sequence of permutations and reversions which transforms B' into B'' such that the virtual vertices appear on the horizontal line in the same order as in B.

The proof is by induction on the number of vertices in the sub-bush forms. If each of the sub-bush forms consists of only one vertex and one virtual vertex, then clearly the proposition is true.

Let v be the smallest vertex of the sub-bush forms. Consider first the case where v is a cutvertex of the sub-bush forms. Then there exist components (sub-bush forms) of B and B', which can be decomposed at v. If the components of B' are not in the same order as in B, then by permuting them, one can put them in the same order as in B. Then, applying the inductive hypothesis to each of the components (sub-bush forms), we can get the sub-bush forms B'' whose virtual vertices are in the same order as in B.

Consider next the case where v is not a cutvertex. Let H be the maximal 2-connected component of the sub-bush forms containing v, and let $u_1, u_2, \ldots, u_k$ be the cutvertices of B contained in H which appear on the outer facial cycle of H in this order. These vertices in B' appear on the outer facial cycle of H in the same or reversed order. If the order is reversed, then reverse H about v in B'. Thus we can put the sub-bush forms of B' rooted with u_i, $1 \leqslant i \leqslant k$, in the same order as in B. Applying the inductive hypothesis to each of the sub-bush forms, we can get B'' whose virtual vertices are in the same order as in B.   Q.E.D.

Booth and Lueker [BL76] showed that the permutations and reversions mentioned in Lemma 3.2 can be found by repeatedly applying the nine transformation rules called the *template matchings* to the PQ-tree. A leaf labelled "$k + 1$" is said to be *pertinent* in a PQ-tree corresponding to B_k. The *pertinent subtree* is the minimal subtree of a PQ-tree containing all the pertinent leaves. A node of a PQ-tree is said to be *full* if all the leaves among its descendants are pertinent. In Fig. 3.6, the template matchings are illustrated. A pattern at the left hand side is a PQ-tree to be transformed and a pattern at the right hand side is a resulting PQ-tree. A full node or subtree is hatched, and a Q-node which roots a pertinent subtree is hatched partially. The template matchings (c) and (e) are for the case when the root of a PQ-tree is a P-node and is the root of a pertinent subtree.

We now present an example of a reduction on a PQ-tree performed by using

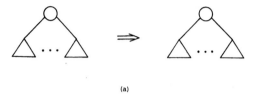

(a)

(b)

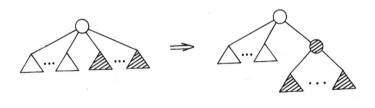

(c)

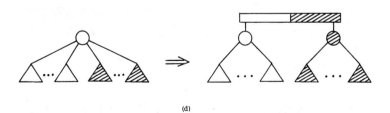

(d)

Fig. 3.6. The template matchings.

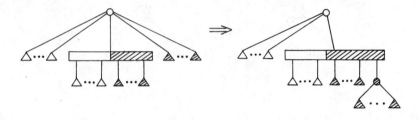

(e)

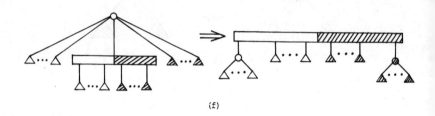

(f)

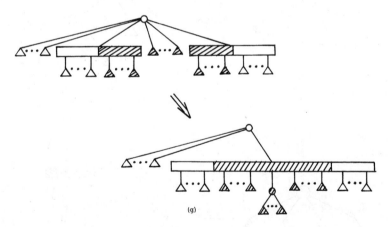

(g)

Fig. 3.6 (continued)

these template matchings. Assume that the *PQ*-tree shown in Fig. 3.7(a) represents a bush form B_{13} of a given graph whose pertinent leaves are labelled "14." Then the *PQ*-trees are transformed as follows. The *PQ*-tree in Fig. 3.7(b)

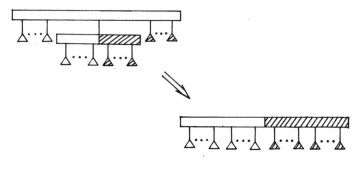

(h)

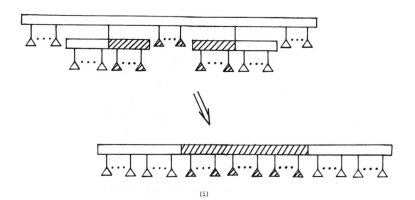

(i)

Fig. 3.6 (continued)

is obtained by applying to P-node P_1 the template matching of type (d) in Fig. 3.6. (Since P_1 has only one full son and one non-full son, P-nodes in the template matching (d) are not created.) Then a template matching of type (h) applied to the PQ-tree in Fig. 3.7(b) at Q_1 produces the PQ-tree in Fig. 3.7(c). A matching of type (h) to Q_2 in Fig. 3.7(c) produces the PQ-tree in Fig. 3.7(d). A matching of type (g) to P_2 in Fig. 3.7(d) produces the PQ-tree in Fig. 3.7(e). In order to construct B_{14}, the PQ-tree in Fig. 3.7(f) is formed from (e) by replacing all the full nodes by a single P-node having sons which correspond to the neighbours of vertex 14 having st-numbers larger than 14. One can easily observe that the next reduction for the vertex 15 will fail, that is, G_{15} is nonplanar.

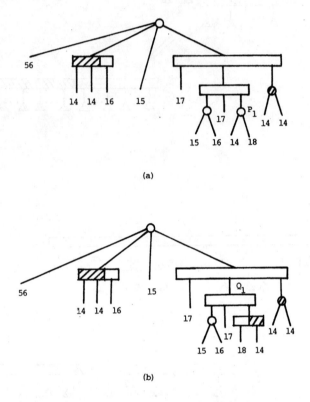

Fig. 3.7. An example of a reduction of a *PQ*-tree.

The following is the outline of the planarity testing algorithm PLANAR using a *PQ*-tree.

procedure PLANAR(G);
    {G is a given graph}
    begin
        assign *st*-numbers to all the vertices of G;
        construct a *PQ*-tree corresponding to G; {a single *P*-node with virtual
        edges incident on source $s = 1$}
        for $v := 2$ *to n*

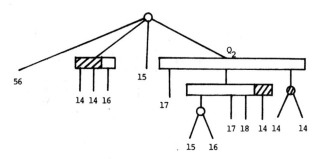

(c)

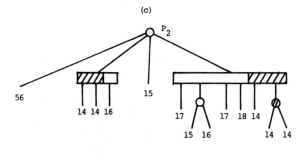

(d)

Fig. 3.7 (continued)

do begin
  {reduction step}
  try to gather all the pertinent leaves by repeatedly applying the template matchings from the leaves to the root of the pertinent subtree;
  if the reduction fails
    then begin
      print out the message "*G* is nonplanar";
      return
    end
  {vertex addition step}
  replace all the full nodes of the *PQ*-tree by a new *P*-node (which corresponds to a cutvertex v in G'_v);
  add to the *PQ*-tree all the neighbours of v larger than v as the sons of the *P*-node
end;

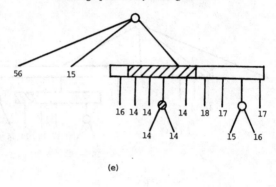

(e)

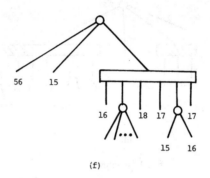

(f)

Fig. 3.7 (continued)

       print out the message "*G* is planar"
end;

Fig. 3.8 illustrates a sequence of bush forms together with the corresponding *PQ*-trees which appeared in the execution of PLANAR for the graph *G* in Fig. 3.1(a). B'_k is a bush form just after the "reduction step" for vertex $k + 1$.

Clearly the time spent by the vertex addition step for v is proportional to the degree of v. Therefore the step spends at most $O(m) = O(n)$ time in total. On the other hand the time spent by the reduction step for v is proportional to the number of pertinent leaves plus the number of unary nodes in the pertinent tree. It is not straightforward to show but was shown by Booth and Lueker that all the reduction steps spend at most $O(n)$ time in total. Thus the algorithm spends at most $O(n)$ time in total.

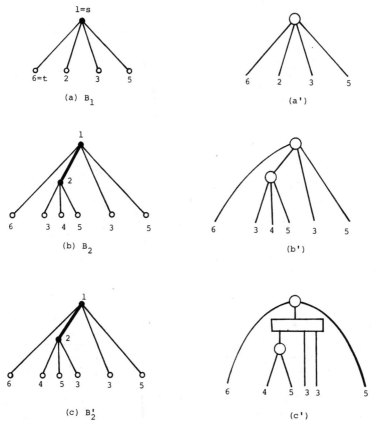

Fig. 3.8. Reduction process of PLANAR for the graph G in Fig. 3.1(a). (a)–(i) bush forms, and (a')–(i') PQ-trees.

3.3. Embedding algorithm

One can easily have the following naive embedding algorithm: first write down the partial embedding of the graph corresponding to B_1; and, with each reduction of the PQ-tree, rewrite (the adjacency lists of) the bush form. Clearly the final bush form is indeed an embedding of the graph. Unfortunately the algorithm spends $O(n^2)$ time, since it takes $O(n)$ time per reduction of the PQ-tree to update the adjacency lists of the bush form.

It this section, we present a linear time embedding algorithm EMBED [CNAO85]. The algorithm EMBED runs in two phases: in the first phase

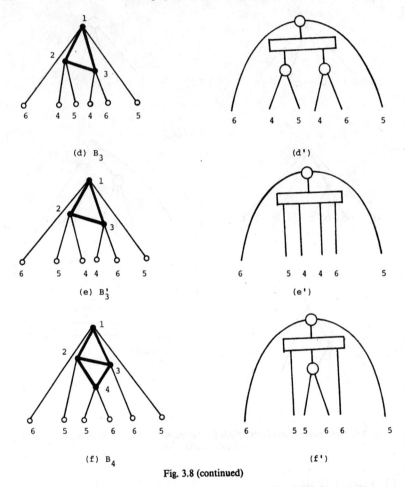

(d) B_3 (d')

(e) B_3' (e')

(f) B_4 (f')

Fig. 3.8 (continued)

EMBED obtains an upward embedding A_u of a planar graph G (by an efficient implementation of the above algorithm); in the second phase EMBED constructs an entire embedding Adj of G from A_u.

3.3.1. Algorithm for extending A_u into Adj

In this subsection we describe an algorithm for the second phase. One can easily observe the following lemma.

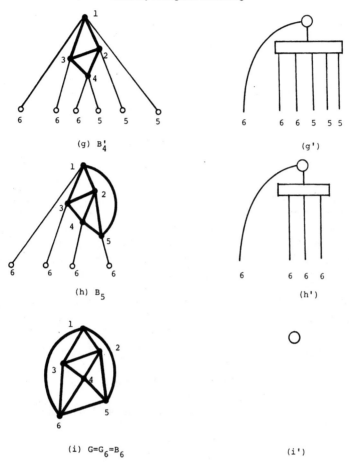

(g) B_4'

(g')

(h) B_5

(h')

(i) $G=G_6=B_6$

(i')

Fig. 3.8 (continued)

Lemma 3.3. *Let* Adj *be an embedding of a planar graph G obtained by the naive algorithm above, and let v be a vertex of G. Then all the neighbours smaller than v are embedded consecutively around v. (See Fig. 3.9.) That is,* Adj(v) *does not contain four neighbours* w_1, w_2, w_3, *and* w_4, *appearing in this order and satisfying* $w_1, w_3 < v$ *and* $w_2, w_4 > v$.

Proof. Immediately follows from Lemmas 3.1 and 3.2.   Q.E.D.

As shown later in Section 3.3.2, EMBED finds an upward embedding A_u of G

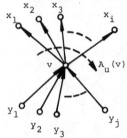

Fig. 3.9. Embedding of the neighbours of v. (Numbers $x_1, x_2, \ldots, x_i$ are all less than v, and $y_1, y_2, \ldots, y_j$ are greater than v.)

such that, for each vertex $v \in V$, the neighbours $x_1, x_2, \ldots, x_i$ smaller than v appear in $A_u(v)$ in this order as indicated by a dotted arrow in Fig. 3.9. That is, the v's neighbour embedded around v counterclockwise next to the top entry x_i of $A_u(v)$ is greater than v. In particular, the top entry of list $A_u(t)$ for sink t is the source $s(=1)$. Now we present the algorithm "ENTIRE-EMBED" for extending such an upward embedding A_u into an embedding Adj of a given graph. The algorithm executes once the depth-first search starting at sink t on a digraph D_u. The algorithm adds vertex y_k to the top of list $A_u(v)$ when the directed edge $\langle y_k, v \rangle$ is searched.

procedure ENTIRE-EMBED;
   begin
      copy the upward embedding A_u to the lists Adj;
      mark every vertex "new";
      $T := \varnothing$; {A DFS-tree T is constructed only for analysis of the algorithm}
      DFS(t)
   end;

procedure DFS(y);
   begin
      mark vertex y "old";
      for each vertex $v \in A_u(y)$
      do begin
        insert vertex "y" to the top of $A_u(v)$;
        if v is marked "new"
          then begin
            add edge $\langle y, v \rangle$ to T;
            DFS(v)
         end

> *end*
> *end*;

We have the following result on the algorithm.

Lemma 3.4. *Let D_u be an upward digraph of a given graph G, and let A_u be an upward embedding of D_u. Then the algorithm ENTIRE-EMBED extends A_u into an embedding* Adj *of G within linear time.*

Proof. Clearly the algorithm terminates within linear time since the algorithm merely executes the depth-first search once. Thus we concentrate on the correctness. The definition of an st-numbering implies that there exists a directed path from t to every vertex. Therefore DFS(t) traverses all the vertices and so all the directed edges of D_u. (This is not necessarily true for an arbitrary digraph.) Hence the final list Adj(v) contains not only the neighbours of v larger than v but also those smaller than v. That is, the final lists Adj are surely adjacency lists of a given (undirected) graph G. Hence we shall prove that all the entries of Adj are stored correctly in clockwise order.

By Lemma 3.3 all v's neighbours $x_1, x_2, \ldots, x_i$ smaller than v appear in $A_u(v)$ in this order. The algorithm first copies list $A_u(v)$ to list Adj(v) and then adds each neighbour y of v larger than v to the top of Adj(v) in the order of the directed edge $\langle y, v \rangle$ being searched. Therefore it suffices to show that directed edges $\langle y_1, v \rangle, \langle y_2, v \rangle, \ldots, \langle y_j, v \rangle$ are searched for in this order. (See Fig. 3.9.) Assume on the contrary that $\langle y_k, v \rangle$ and $\langle y_l, v \rangle$ are searched for in this order although $k > l$. Let P_k be the path from t to y_k, and let P_l be the path from t to y_l in the DFS-tree T. (See Fig. 3.10.) Let z be the vertex at which path P_l leaves P_k, and let $\langle z, y_k' \rangle \in P_k$ and $\langle z, y_l' \rangle \in P_l$. Thus the vertex y_k' precedes y_l' in $A_u(z)$. Moreover the subpaths $P_k' = z \cdot y_k' \cdots y_k$ of P_k and $P_l' = z \cdot y_l' \cdots y_l$ of P_l have no common vertices other than z. Therefore the two paths P_k' and P_l' together with two edges (y_l, v) and (y_k, v) form a cycle C. All the vertices of $A_u(v)$ must lie in the interior of the cycle; otherwise Lemma 3.3 would be violated. Since source $s (= 1)$ is located on the boundary of the exterior face, the vertex v is not s. By the definition of an st-numbering, the DFS-tree T contains a descending path P from v to s, all the vertices of which are smaller than or equal to v. Since s lies in the exterior of the cycle C, P must intersect the cycle C. However all the vertices of C are larger than or equal to v. This is a contradiction.　　Q.E.D.

3.3.2. *Algorithm for constructing A_u*

In this subsection we give an algorithm for constructing A_u. One can easily

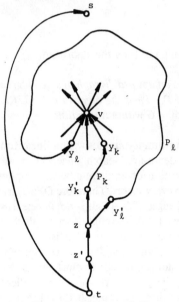

Fig. 3.10. Illustration for the proof of Lemma 3.4.

obtain list $A_u(v)$ or its reversion by scanning the leaves labelled "v" in the vertex addition step for v. If $A_u(v)$ is correctly determined in the step, then, counting the number of subsequent v's reversions, one can correct the direction of $A_u(v)$ simply by reversing $A_u(v)$ if the number is odd. However a naive counting algorithm takes $O(n^2)$ time. Moreover, the information on "v" may disappear from the PQ-tree. Thus an appropriate device is necessary to trace v's reversions.

We first show how to scan all the leaves labelled "v". The root r of the pertinent subtree can be found by the "bubble up" procedure [BL76]. Let $b_1, b_2, \ldots, b_m$ be the maximal sequence of full brothers that are sons of r. (See Fig. 3.11(a).) To obtain $A_u(v)$, we scan the subtree rooted at b_i by the depth-first search for $i = 1, 2, \ldots, m$ in this order. In a schematic illustration of a PQ-tree, one can easily recognize the direction of the maximal sequence, that is, whether $b_1, b_2, \ldots, b_m$ are in left-to-right or right-to-left order. However in the data structure of a PQ-tree, a Q-node is doubly linked only with the endmost sons, and a son of a Q-node has pointers only to the immediate brothers [BL76]. Therefore we must traverse sons of a Q-node from a full son to one of the endmost sons, and then check the direction of the sequence by using the pointer between the endmost son and the Q-node. Thus such a straightforward method requires $O(n)$ time to determine the direction of the sequence, that is, to know whether the constructed list is either $A_u(v)$ or its reversion.

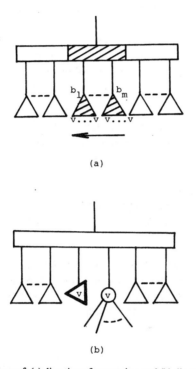

(a)

(b)

Fig. 3.11. Illustrations of: (a) direction of a scanning, and (b) direction indicator "v".

The algorithm does not determine the direction of $A_u(v)$ at the vertex addition step for v, but adds a new special node to the PQ-tree as one of r's sons at an arbitrary position among them. The new node is called a "direction indicator", also labelled "v", and depicted by a triangle, as illustrated in Fig. 3.11(b). The indicator "v" plays two roles. The first is to trace the subsequent reversions of v. The indicator will be reversed with each reversion of its father. (No physical action is taken in the indicator's reversion — it is only done implicitly.) The second is to bear the relative direction of node "v" to its brothers. When the rightmost or leftmost brother of "v" is subsequently scanned together with the indicator "v", the direction of the constructed $A_u(v)$ is known and so is corrected if necessary.

In the template matching algorithm, we ignore the presence of the direction indicators. When we access an immediate brother b of a node v, we skip the direction indicators between v and b if any. When we change pointers of a PQ-tree in a reduction step, we treat a direction indicator as a usual node of a PQ-tree. Note that all the direction indicators in a PQ-tree are necessarily leaves: none of the direction indicators has a son. Now we redefine a node to be *full* if all the leaves of its descendants which are not indicators are labelled "v".

Thus we modify the vertex addition step in PLANAR as follows.

{vertex addition step}
begin
    let $l_1, l_2, \ldots, l_j$ be the leaves labelled "v", and let $f_1, f_2, \ldots, f_k$ be the direction indicators scanned (using the DFS procedure just described) in this order {it is not necessary to recognize here whether $l_1, l_2, \ldots, l_j$ are in left-to-right order};
    $A_u(v) := \{l_1, l_2, \ldots, l_j\}$;
    if root r of the pertinent subtree is not full
        {the subtree has a leaf which is not an indicator and not labelled "v"}
    then {the root r is a Q-node}
        begin
            add an indicator "v" directed from l_j to l_1 to the PQ-tree as a son of the Q-node r {at an arbitrary position among the sons};
            add the direction indicators $f_1, f_2, \ldots, f_k$ as sons of the Q-node r {at arbitrary positions among the sons}
        end
    else
        begin {the pertinent subtree corresponds to a reversible component in an embedding of G, that is, the cutvertex of G corresponding to root r forms a "separation pair" with vertex v. Therefore we may assume that $A_u(v)$ is correctly in clockwise order.}
    delete $f_1, f_2, \ldots, f_k$ from the PQ-tree;
    for $i := 1$ *to* k
        do if indicator f_i is directed from l_1 to l_k
            then reverse the adjacency list $A_u(f_i)$;
            {The order of $A_u(f_i)$ is corrected with the assumption that $A_u(v)$ is in clockwise order.}
    end;
    if root r is not full
    then replace all the full sons of r by a P-node {which corrresponds to a cutvertex v of G'_v}
    else replace the pertinent subtree by a P-node;

add all the virtual vertices adjacent to v (i.e. all neighbours of v in G greater than v) to the PQ-tree as the sons of the P-node
end;

We call this revised algorithm UPWARD-EMBED, on which we have the following result.

Lemma 3.5. *The algorithm UPWARD-EMBED obtains an upward embedding A_u of a given planar graph G.*

Proof. Let $v \in V$. Clearly the list $A_u(v)$ obtained by UPWARD-EMBED contains all the neighbours of v smaller than v. Furthermore these vertices appear in either clockwise or counterclockwise order around v. Therefore we shall show that the vertices in each $A_u(v)$ appear in clockwise order. It suffices to consider the following two cases.

Case 1: the direction indicator "v" is not added to the PQ-tree. The leaves of the pertinent subtree which are not indicators are all labelled "v" at the vertex addition step for v. Such a pertinent subtree corresponds to a reversible component in a plane embedding of G. (See Fig. 3.12.) Therefore one may assume that the vertices in $A_u(v)$ appear in clockwise order even in the final embedding.

Case 2: the direction indicator "v" is added to the PQ-tree. When the algorithm terminates, the PQ-tree consists of exactly one isolated P-node, and hence has no direction indicators in particular. That is, every indicator will be eventually deleted. Therefore one can assume that the indicator "v" is deleted in the vertex addition step for a vertex $w(>v)$. The direction indicator "v" follows reversions of the Q-node which is the father of node "v" as long as v remains in a PQ-tree. Therefore if the direction of indicator "v" is opposite relative to the scanning of the leaves $l_1, l_2, \ldots, l_j$ labelled "w", then either the order (clockwise or counterclockwise) of $A_u(v)$ is the same as $A_u(w)$ and vertex v is reversed an odd number of times, or the order of $A_u(v)$ is opposite to that of $A_u(w)$ and the vertex v is reversed an even number of times. In either case, we can correct adjacency list $A_u(v)$ simply by reversing it. Since the pertinent subtree for w corresponds to a reversible component of G, the direction indicator "w" is not added to the PQ-tree. Hence the adjacency lists $A_u(v)$ and $A_u(w)$ are never reversed after the vertex addition step for w. Thus $A_u(v)$ remains correctly in clockwise order. Q.E.D.

However, algorithm UPWARD-EMBED, as it is, requires $O(n^2)$ time since it may scan the same indicator many times, say $O(n)$ times. Thus we shall refine the algorithm so that it runs in $O(n)$ time.

Now consider the role of a direction indicator in detail. Assume that root r of a pertinent subtree is not full, and define indicators "v" and $f_1, f_2, \ldots, f_k$ as in the algorithm. After the direction indicator "v" is added to a PQ-tree, indicators "v" and $f_1, f_2, \ldots, f_k$ are reversed all together. Therefore it suffices to remember the directions of $f_1, f_2, \ldots, f_k$ relative to that of "v". Thus we delete the indicators $f_1, f_2, \ldots, f_k$ from the PQ-tree and store them in $A_u(v)$ together with vertices $l_1, l_2, \ldots, l_j$. Once the correct order of adjacency list $A_u(v)$ is known, we can easily correct the orders of adjacency lists $A_u(f_i)$, $1 \le i \le k$,

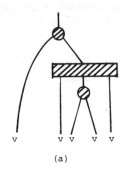

(a)

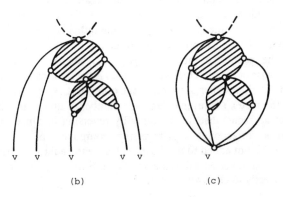

(b)   (c)

Fig. 3.12. Reversible component. (a) pertinent subtree, (b) B_{v-1}, and (c) G_v.

simply by checking the direction of indicator f_i in $A_u(v)$. We execute such a correction for each v, $v = n, n - 1, \ldots, 1$, in this order.

The following is the algorithm UPWARD-EMBED refined as above. Fig. 3.13 illustrates an UPWARD-EMBED applied to the graph in Fig. 3.1(a).

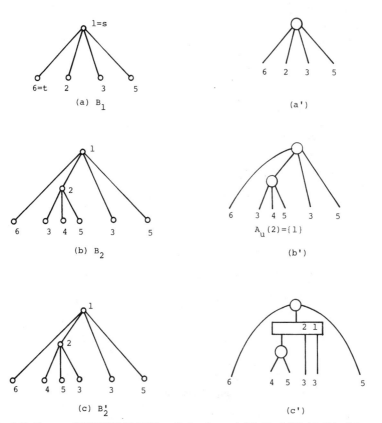

Fig. 3.13. Process of UPWARD-EMBED applied to the graph G in Fig. 3.1(a). (a)–(i) bush forms, and (a′)–(i′) PQ-trees and lists A_u, and (j) corrected lists A_u.

procedure UPWARD-EMBED(G);
   {G is a given planar graph}
   begin
      assign *st*-numbers to all the vertices of G;
      construct a PQ-tree corresponding to G'_1;

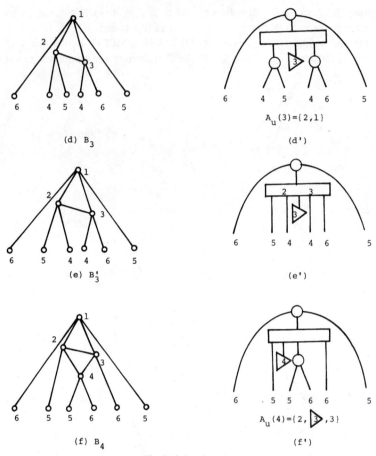

(d) B_3

$A_u(3) = \{2, 1\}$

(d')

(e) B_3'

(e')

(f) B_4

$A_u(4) = \{2, \triangleright, 3\}$

(f')

Fig. 3.13 (continued)

for $v := 2$ *to* n
   do begin
      {reduction step}
      apply the template matchings to the *PQ*-tree, ignoring the direction
      indicators in it, so that the leaves labelled "v" occupy consecutive
      positions;
      {vertex addition step}
      let $l_1, l_2, \ldots, l_k$ be the leaves labelled "v" and direction indicators
      scanned in this order;

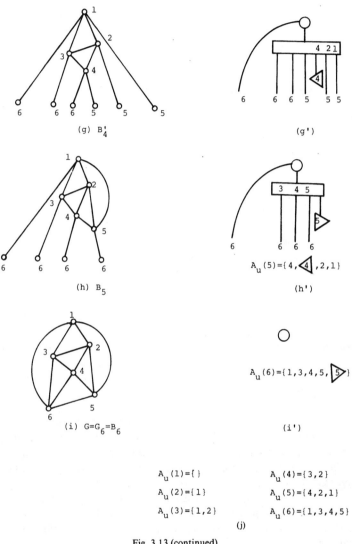

$$A_u(5) = \{4, \blacktriangleleft, 2, 1\}$$

(h')

$$A_u(6) = \{1, 3, 4, 5, \triangleright\}$$

(i')

$A_u(1) = \{\ \}$	$A_u(4) = \{3, 2\}$
$A_u(2) = \{1\}$	$A_u(5) = \{4, 2, 1\}$
$A_u(3) = \{1, 2\}$	$A_u(6) = \{1, 3, 4, 5\}$

(j)

Fig. 3.13 (continued)

delete $l_1, l_2, \ldots, l_k$ from the *PQ*-tree and store them in $A_u(v)$;
if root r of the pertinent subtree is not full
    then
    begin

add an indicator "v", directed from l_k to l_1, to the PQ-tree as a son
of root r at an arbitrary position among the sons;
replace all the full sons of r by a new P-node
end
else
replace the pertinent subtree by a new P-node;
add to the PQ-tree all the virtual vertices adjacent to v as the sons of the
P-node
end;
{correction step}
for $v := n$ *down to* 1
  do for each element x in $A_u(v)$
    do if x is a direction indicator
      then begin
        delete x from $A_u(v)$;
        let w be the label of x;
        if the direction of indicator x is
          opposite to that of $A_u(v)$
        then reverse list $A_u(w)$;
      end
end;

We have the following result on the revised UPWARD-EMBED above.

Lemma 3.6. *Algorithm UPWARD-EMBED obtains the upward embedding A_u
of a given planar graph within linear time.*

Proof. Noting the role of a direction indicator, one can easily verify the
correctness of the algorithm. Therefore we consider the time required by the
algorithm. At most $O(n)$ direction indicators are generated during an execution of the algorithm. A direction indicator scanned in a reduction step will
be necessarily deleted from a PQ-tree in a succeeding vertex addition step.
Therefore each direction indicator is scanned at most once. Thus UPWARD-
EMBED requires at most $O(n)$ time in addition to the time required by the
linear testing algorithm PLANAR. Therefore UPWARD-EMBED runs in
linear time.   Q.E.D.

The following is the entire algorithm EMBED for embedding a planar graph.

procedure EMBED(G);
  begin

UPWARD-EMBED; {phase 1}
ENTIRE-EMBED {phase 2}
end;

We have the following Theorem 3.2 from Lemmas 3.1–3.6.

Theorem 3.2. *Algorithm EMBED obtains a plane embedding* Adj *of a given planar graph within linear time.*

CHAPTER 4

DRAWING PLANAR GRAPHS

4.1. Introduction

The problem of drawing a planar graph often arises in applications, including the Design Automation of VLSI circuits. In this chapter we discuss algorithms to draw a plane graph by straight line segments. Some planar graphs can be drawn in such a way that all the face boundaries are convex polygons. Such a drawing is called a *convex* one. Although not every planar graph has a convex drawing, Tutte [Tut60] proved that every 3-connected planar graph has a convex drawing, and established a necessary and sufficient condition for a planar graph to have a convex drawing. Furthermore he gave a "barycentric mapping" method for finding a convex drawing, which requires solving a system of $O(n)$ linear equations [Tut63]. The system of equations can be solved in $O(n^3)$ time and $O(n^2)$ space using the ordinary Gaussian elimination method, or in $O(n^{1.5})$ time and $O(n \log n)$ space using the sparse Gaussian elimination method [LRT79]. Thus the barycentric mapping method leads to an $O(n^{1.5})$ time convex drawing algorithm.

In this Chapter we present two linear algorithms for the convex drawing problem of planar graphs: drawing and testing algorithms. These are given by Chiba, Yamanouchi and Nishizeki [CYN84]. The former draws a given planar graph G convex if possible: it extends a given convex polygonal drawing of an outer facial cycle of G into a convex drawing of G. Thomassen gave a short proof of Tutte's result [Tho80]. The drawing algorithm is based on the proof. The latter algorithm tests the possibility. That is, it determines whether a given planar graph has a convex drawing or not. Chiba et al. [CYN84] showed that the convex testing of a graph G can be reduced to the planarity testing of a certain graph obtained from G. The testing algorithm employs this fact together with two linear algorithms: one for testing the planarity of a graph (Chapter 3); and the other for dividing a graph into 3-connected components

[HT74]. In Section 4.4 we give a computational example: a convex drawing of an input planar graph.

4.2. Convex drawing

The purpose of this section is to present a linear convex drawing algorithm of planar graphs. We first define some terms and illustrative examples, then present Thomassen's result on the convex drawing, and finally give the algorithm.

The vertex set of a graph G is often denoted by $V(G)$. We consider only a drawing of a 2-connected graph. A path joining vertices x and y is called an *x-y path*. A *convex drawing* of a planar graph is a representation of the graph on the plane such that all edges are drawn by straight line segments without any crossing and that all the face boundaries are convex polygons. Since all the edges are drawn by straight lines, a convex drawing of a plane graph is uniquely determined only by the positions of the vertices.

Clearly not every 2-connected planar graph has a convex drawing. For example, the 2-connected planar graph depicted in Fig. 4.1(a) cannot be drawn convex even if any facial cycle is chosen as an outer cycle. Next consider the graph G in Fig. 4.1(b). In this case it depends on the facial cycle chosen as an outer cycle whether G can be drawn convex or not. If the outer facial cycle S is 1–2–3–4–1, G cannot be drawn convex for any polygonal drawing S^* of S, as shown in Fig. 4.1(c). On the other hand, G can be drawn convex if the facial cycle $S = 1$–2–3–4–5–6–1 is chosen as an outer cycle and moreover S is drawn as a convex polygon S^* such that vertices 1, 2, 3, 4, and 6 are the apices (i.e. geometric vertices) of S^*, as shown in Fig. 4.1(d). However, if S^* is a convex polygon with the apices 1, 2, 3, and 4, then G cannot be drawn convex as shown in Fig. 4.1(e). Thus we define: a convex polygonal drawing (for short, a convex polygon) S^* of a facial cycle S of a graph G is *extendible* if there exists a convex drawing of G having S^* as the outer polygon; a facial cycle S is *extendible* if S has an extendible convex polygon S^*.

Tutte established a necessary and sufficient condition for a convex polygon to be extendible [Tut60]. The following lemma obtained by Thomassen [Tho80] is slightly more general than his result.

Lemma 4.1. [*Tho*80] *Let G be a 2-connected plane graph with the outer facial cycle S, and let S^* be a convex polygon of S. Let $P_1, P_2, \ldots, P_k$ be the paths in S, each corresponding to a side of S^*. (Thus S^* is a k-gon. Not every vertex of the cycle S is an apex of the polygon S^*.) Then S^* is extendible if and only if Condition I below holds.*

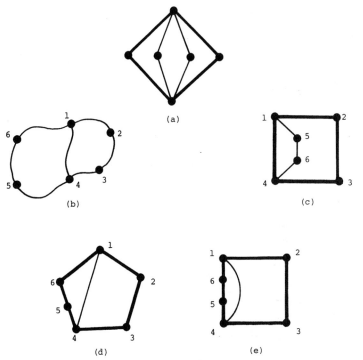

Fig. 4.1. Examples of drawings.

Condition I

(a) For each vertex v of $G - V(S)$ having degree at least three in G, there exist three paths disjoint except v, each joining v and a vertex of S;

(b) $G - V(S)$ has no connected component C such that all the vertices on S adjacent to vertices in C lie on a single path P_i; and no two vertices in each P_i are joined by an edge not in S; and

(c) any cycle of G which has no edge in common with S has at least three vertices of degree ≥ 3 in G.

Proof. The situations violating Conditions I (a)–(c) are illustrated in Fig. 4.2. One can easily observe the necessity of Condition I. On the other hand the sufficiency is implied by the algorithm, to be described soon, which always finds a convex drawing of a plane graph satisfying the condition.   Q.E.D.

Suppose that a 2-connected plane graph G is given together with an

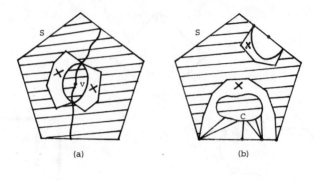

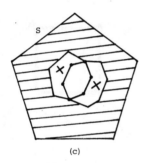

Fig. 4.2. Situations violating Conditions I(a)–(c).

extendible convex polygon S^* of the outer facial cycle S. The linear convex drawing algorithm extends S^* into a convex drawing of G in linear time. For simplicity we assume that every vertex not on S has degree three or more in G. Otherwise, replace each maximal induced path not on S by a single edge joining its ends (the resulting simple graph G' satisfies Condition I); then find a convex drawing of G'; and finally subdivide the edges substituting the paths to complete a convex drawing of G.

The outline of the drawing algorithm is as follows. We reduce the convex drawing of G to those of several subgraphs of G as follows: delete from G an arbitrary apex v of S^* together with the edges incident to v; divide the resulting graph $G' = G - v$ into the blocks $B_1, B_2, \ldots, B_p$, $p \geqslant 1$ (see Fig. 4.3); determine a convex polygon S_i^* of the outer facial cycle S_i of each B_i so that B_i with S_i^* satisfies Condition I; and recursively apply the algorithm to each B_i with S_i^*

to determine the positions of vertices not in S_i. The detail of the algorithm is as follows.

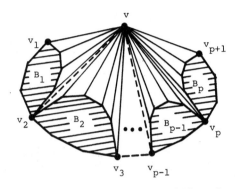

Fig. 4.3. Reduction of the convex drawing of G into subproblems.

Algorithm CONVEX-DRAW(G, S, S^*)

Step 1. Assume that G has at least four vertices; otherwise a convex drawing of G has been obtained. Select an arbitrary apex v of S^*, and let $G' := G - v$. Divide the plane graph G' into the blocks B_i ($1 \leqslant i \leqslant p$). Let v_1 and v_{p+1} be the two vertices on S adjacent to v, and let v_i, $2 \leqslant i \leqslant p$, be the cutvertices of G' such that $v_1 \in V(B_1)$, $v_{p+1} \in V(B_p)$ and $v_i = V(B_{i-1}) \cap V(B_i)$. (See Fig. 4.3.) {Every v_i, $1 \leqslant i \leqslant p + 1$, is necessarily on S since the extendible S^* with G satisfies Condition I and G has no vertex of degree two not on S.}

Step 2. Draw each block B_i convex by applying the following procedures.
  Step 2.1. Determine a convex polygon S_i^* of the outer facial cycle S_i of B_i as follows. {Since the positions of the vertices in $V(S_i) \cap V(S)$ have been already determined on S^*, it remains to determine the positions of the vertices in $V(S_i) - V(S)$.} Locate the vertices in $V(S_i) - V(S)$ in the interior of the triangle $v \cdot v_i \cdot v_{i+1}$ in such a way that the vertices adjacent to v are apices of a convex polygon S_i^* and the others are on the straight line segments of S_i^*.
  Step 2.2. Recursively call the procedure CONVEX-DRAW(B_i, S_i, S_i^*) to extend S_i^* to a convex drawing of B_i. {Note that the convex polygon S_i^* with B_i satisfies Condition I and B_i has no vertex of degree two not on S_i.}

We have the following result on the algorithm.

Theorem 4.1. *Let G be a 2-connected plane graph with the outer facial cycle S, and let S^* be an extendible convex polygon of S. Then the algorithm CONVEX-DRAW extends S^* into a convex drawing of G, and uses linear time and space.*

Proof. The boundaries of all the inner faces containing the apex v are triangles, and hence they are convex polygons. Therefore, in order to prove inductively the correctness of the algorithm, it suffices to observe that every block B_i with S_i^* satisfies Condition I. If B_i violated Condition I(a) or (c) then G would do the same one, while if B_i violated (b) then G would do either (a), (b) or (c). Thus we shall establish the claims on time and space.

As a data structure to represent a plane graph $G = (V, E)$, we use doubly linked adjacency lists, in each of which the edges adjacent to a vertex are stored in the order of the plane embedding, clockwise around the vertex. The two copies of each edge (v, w) in the adjacency lists of v and w are linked to each other so that one can be accessed directly from the other. Given an edge e, one can directly access the edge clockwise next to e around an end vertex of e. Clearly this data structure requires linear space.

Let P be the $v_1 - v_{p+1}$ path in the outer facial cycle S' of $G - v$ which newly appears on S'. Traversing P, one can easily (1) find the cutvertices v_i, $2 \leqslant i \leqslant p$, which appear on both S and S', (2) obtain the outer facial cycle S_i of B_i as the union of the traversed $v_i - v_{i+1}$ path and the $v_{i+1} - v_i$ path on S, and (3) decide the positions of the vertices of S_i as specified in Step 2.1. Thus, we can implement the algorithm so that the required time, exclusive of recursive calls to itself, is proportional to the number of the traversed edges in P, that is, the edges that newly appear on the boundaries of the outer facial cycles. Since every edge appears on a boundary of an outer facial cycle at most once, the number of edges traversed during an execution of DRAW is at most $|E|$ in total. Thus DRAW runs in linear time. Q.E.D.

4.3. Convex testing

In this section, we present a linear convex testing algorithm of [CYN84] which determines whether a given 2-connected planar graph has a convex drawing.

It is rather easy to design a linear algorithm which only determines whether a given convex polygon of a particular outer facial cycle is extendible, that is, satisfies Condition I. However a planar graph G may have an exponential number of facial cycles, so it is impractical to test all the facial cycles of a graph one by one through such an algorithm. Note that the plane embedding of G is not always unique unless G is 3-connected (Theorem 1.1). Thus we shall modify Condition I in Lemma 4.1 into a form suitable for our purpose, which is represented in terms of "3-connected components." One may easily notice that the existence of a convex drawing of a graph G heavily depends on the structure of "3-connected components" of G.

This section is organized as follows. In Section 4.3.1 we give definitions of 3-connected components and separation pairs. In Section 4.3.2 we express Condition I in terms of 3-connected components. Section 4.3.3 gives a linear convex testing algorithm.

4.3.1. Definitions

In this subsection "graph" means the so-called multigraph, so some terms defined in Chapter 2 shall be redefined. A pair $\{x, y\}$ of vertices of a 2-connected graph $G = (V, E)$ is a *separation pair* if there exist two subgraphs $G_1' = (V_1, E_1')$ and $G_2' = (V_2, E_2')$ satisfying the following conditions (a) and (b):

(a) $\qquad V = V_1 \cup V_2, \quad V_1 \cap V_2 = \{x, y\};$

(b) $\qquad E = E_1' \cup E_2', \quad E_1' \cap E_2' = \varnothing, \quad |E_1'| \geqslant 2, \quad |E_2'| \geqslant 2.$

A 2-connected graph G is said to be *3-connected* if G has no separation pair. For a separation pair $\{x, y\}$, $G_1 = (V_1, E_1' + (x, y))$ and $G_2 = (V_2, E_2' + (x, y))$ are called *split graphs* of G. The new edges (x, y) added to G_1 and G_2 are called *virtual edges*. Even if G has no multiple edges, G_1 or G_2 may have. Dividing a graph G into two split graphs G_1 and G_2 is called *splitting*. Reassembling the two split graphs G_1 and G_2 into G is called *merging*. Merging is the inverse of splitting. Suppose a graph G is split, the split graphs are split, and so on, until no more splits are possible (each remaining graph is 3-connected). The graphs constructed in this way are called the *split components* of G. The split components of a graph G are of three types: *triple bonds* (i.e. a set of three multiple edges), *triangles*, and 3-connected graphs. The *3-connected components* of G are obtained from the split components of G by merging triple bonds into a bond and triangles into a ring, as far as possible. Here a *bond* is a set of multiple edges, and a *ring* is a cycle. The split components of graph G are not necessarily unique, but the 3-connected components of G are unique.

We illustrate the decompositions of a 2-connected graph in Fig. 4.4. The graph G depicted in Fig. 4.4(a) has six separation pairs $\{1, 2\}$, $\{1, 3\}$, $\{2, 3\}$, $\{2, 7\}$, $\{3, 6\}$, and $\{4, 5\}$. The graph G is decomposed into nine split components as shown in Fig. 4.4(b), and into seven 3-connected components

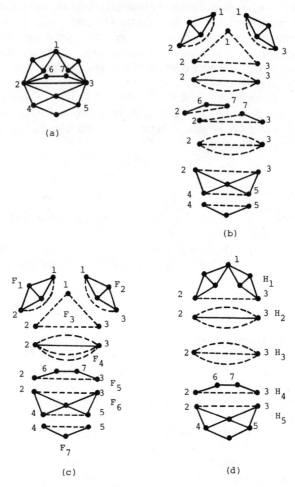

Fig. 4.4. Decompositions of a graph, where virtual edges are written by dashed lines:
(a) a 2-connected graph G; (b) split components of G; (c) 3-connected components of G; and
(d) {2, 3}-split components of G with one exception H_3.

$F_1, F_2, \ldots, F_7$ as shown in Fig. 4.4(c). The components F_1, F_2 and F_6 are
3-connected graphs; F_3, F_5 and F_7 are rings; and F_4 is a bond.

Next we introduce new terms. Suppose that $\{x, y\}$ is a separation pair of a
graph G and that G is split at $\{x, y\}$, the split graphs are split, and so on, until
no more splits are possible at $\{x, y\}$ (the remaining graphs are not necessarily

3-connected). A graph constructed in this way is called an $\{x, y\}$-*split component* of G if it has at least one real (i.e. non-virtual) edge. In Fig. 4.4(d), the components H_1, H_2, H_4, and H_5 are the $\{2, 3\}$-split components.

A separation pair $\{x, y\}$ is *prime* if x and y are the end vertices of a virtual edge contained in a 3-connected component. In other words, a separation pair $\{x, y\}$ is prime if there exist two subgraphs G_1' and G_2' such that G_1' and G_2' satisfy the conditions (a) and (b) aforementioned and G_1' or G_2' is 2-connected or is a subdivision of an edge joining two vertices of degree three or more. As known from Fig. 4.4(c), separation pairs $\{1, 2\}$, $\{1, 3\}$, $\{2, 3\}$, and $\{4, 5\}$ are prime but $\{2, 7\}$ and $\{3, 6\}$ are not.

In some cases it can be easily established from the number of $\{x, y\}$-split components that a graph can never be drawn convex. A *forbidden separation pair* $\{x, y\}$ is a prime separation pair which has either (i) at least four $\{x, y\}$-split components or (ii) three $\{x, y\}$-split components none of which is a ring or bond. Note that an $\{x, y\}$-split component corresponds to an edge (x, y) of G if it is a bond, and corresponds to a subdivision of an edge (x, y) if it is a ring. The graph in Fig. 4.4(a) has exactly one forbidden separation pair $\{2, 3\}$. It will be shown later in Section 4.3.2 that if a planar graph G has a forbidden separation pair then G can never be drawn convex: G has no extendible facial cycle.

On the other hand the converse of the fact above is not true. In order to be more precise, we need one more term. A *critical separation pair* $\{x, y\}$ is a prime separation pair which has either (i) three $\{x, y\}$-split components including a ring or a bond, or (ii) two $\{x, y\}$-split components neither of which is a ring. In the graph of Fig. 4.4(a), prime separation pairs $\{1, 2\}$ and $\{1, 3\}$ are critical, but $\{4, 5\}$ is neither forbidden nor critical. When G has no forbidden separation pair, two cases occur: if G has no critical separation pair either, then G is a subdivision of a 3-connected graph, and so every facial cycle of G is extendible; otherwise, that is, if G has critical separation pairs, a facial cycle S of G may or may not be extendible, depending on the interaction of S and critical separation pairs. (The detailed criterion, called Condition II, will be given in subsection 4.3.2.)

4.3.2. Condition II

We now give a condition suitable for the testing algorithm, which is equivalent to Condition I under a restriction that S^* is *strict*, that is, every vertex of S is an apex of S^*.

Lemma 4.2. [CYN84] *Let* $G = (V, E)$ *be a 2-connected plane graph with the* *outer facial cycle S, and let S* be a strict convex polygon of S. Then S* is extendible if and only if G and S satisfies the following Condition II.*

Condition II
(a) *G* has no forbidden separation pair;
(b) For each critical separation pair $\{x, y\}$ of *G* there exists at most one $\{x, y\}$-split component having no edge of *S*. Moreover, such an $\{x, y\}$-split component is either a bond if $(x, y) \in E$ or a ring otherwise.

Proof. We shall show that Condition II is equivalent to Condition I under the restriction that *S** is strict.

Condition I implies Condition II: Let $\{x, y\}$ be a prime separation pair of *G*, and let $H_1, H_2, \ldots, H_k$ be the $\{x, y\}$-split components having no edges of *S*. Then we can show that $k = 0$ or 1 and that if $k = 1$ then H_1 is either a ring or a bond, as follows. First suppose that one of the $\{x, y\}$-split components, say H_1, is neither a ring nor a bond, then H_1 has a vertex v ($\neq x, y$) of degree three or more. Clearly there exist no three paths disjoint except v, each joining v and a vertex of *S*, since such a path must contain either *x* or *y*. This contradicts Condition I(a). Thus every H_i must be either a ring or a bond. Next suppose that $k \geq 2$, then H_1 together with H_2 forms a cycle in *G* which has no edge in common with *S*. The cycle has exactly two vertices *x* and *y* of degree ≥ 3 in *G*, contrary to Condition I(c).

Since at most two $\{x, y\}$-split components contain edges of *S*, there are at most three $\{x, y\}$-split components and moreover one of them is a bond or a ring if there are three. Thus $\{x, y\}$ is not forbidden.

Let $\{x, y\}$ be critical and $k = 1$. Consider first the case $(x, y) \notin E$. Then clearly H_1 is not a bond, so H_1 must be a ring. Consider next the case $(x, y) \in E$. If H_1 was a ring, then edge (x, y) would be in *S*, and so the *x–y* path in H_1, which is a connected component of $G - V(S)$, would be adjacent only with the vertices *x* and *y* in *S*, contradicting Condition I(b). Note that edge (x, y) is a side of *S**. Thus H_1 must be a bond in this case.

Condition II implies Condition I: First suppose that *G* has a vertex *v* of degree three or more, violating Condition I(a). Then, using Menger's theorem [Har72, p. 47], one can easily show that there exists a prime separation pair $\{x, y\}$ such that one of the $\{x, y\}$-split components H_i contains *v* and has no edge of *S*. Since H_i contains a vertex *v* of degree three or more, H_i is neither a ring nor a bond, contradicting Condition II(b).

Next suppose that Condition I(b) is violated. Then there exists a connected component *C* of $G - V(S)$ such that only the vertices *x* and *y* of an edge (x, y) on *S* are adjacent with vertices in *C*, because *G* has no multiple edges and *S** is strict. Therefore $\{x, y\}$ is a prime separation pair, and clearly the $\{x, y\}$-split component containing C has no edge in common with S and is not a bond since G is simple. This contradicts Condition II(b).

Finally suppose that there exists a cycle Z in G violating Condition I(c). Since G is 2-connected, Z has exactly two vertices x and y of degree ≥ 3. Clearly $\{x, y\}$ is a prime separation pair. If $(x, y) \in E$, then an $\{x, y\}$-split component having no edges of S is a ring. Otherwise, there are two $\{x, y\}$-split components having no edges of S. Either case contradicts Condition II. Q.E.D.

It should be noted that Condition II does not depend on the drawing S^* of S at all. One may suspect that the restriction on S^* loses generality: there would be an extendible convex polygon of S even if there is no extendible strict S^*. However the following lemma dispels this suspicion.

Lemma 4.3. *Assume that G is a 2-connected plane graph with the outer facial cycle S, and that S has an extendible convex polygon. Then every strict convex polygon of S is extendible.*

Proof. Immediately follows from Condition I.   Q.E.D.

Lemmas 4.2 and 4.3 immediately lead to the following theorem.

Theorem 4.2. *A facial cycle S of a 2-connected planar graph G is extendible if and only if S and G satisfy Condition II.*

4.3.3. Testing algorithm

Condition II is more suitable for testing than Condition I. In this subsection, we show that the convex testing, i.e., checking Condition II, can be reduced to the planarity testing of a certain graph.

Theorem 4.2 immediately yields the following corollaries.

Corollary 4.1. *A 2-connected planar graph G has no convex drawing if G has a forbidden separation pair.*

Corollary 4.2. *A 2-connected planar graph G has a convex drawing for any facial cycle of G if G has no forbidden or critical separation pairs.*

Corollary 4.3. *Every 3-connected planar graph G has a convex drawing for any facial cycle of G.*

Corollary 4.4. *Every plane graph G has a straight-line drawing.*

Proof. Augment G to a maximal plane graph G' by adding edges. Since G' is 3-connected, G' has a convex drawing, and hence G has a straight-line drawing.
Q.E.D.

Corollary 4.5. *If a facial cycle S of a 2-connected planar graph G satisfies Condition II, then S contains every vertex of critical separation pairs of G.*

Proof. Assume that $\{x, y\}$ is a critical separation pair of G and that S does not contain a vertex x of the pair $\{x, y\}$. Then exactly one $\{x, y\}$-split component contains all the edges of S. On the other hand, Condition II implies that there exists exactly one $\{x, y\}$-split component not containing edges of S, and it must be a ring or a bond. Thus there are exactly two $\{x, y\}$-split components, one of which is a ring or a bond. Then $\{x, y\}$ could not be critical, contrary to the assumption.
Q.E.D.

We will show later in Theorem 4.3 that the converse of Corollary 4.5 is also true in a certain sense. Before presenting Theorem 4.3 we need the following two lemmas.

Lemma 4.4. *Suppose that a 2-connected planar graph G has no forbidden separation pair and has exactly one critical separation pair. Then G has a convex drawing.*

Proof. Let $\{x, y\}$ be the critical separation pair. By the definition of a critical separation pair, one can easily observe that G is one of the seven types of Fig. 4.5, in which a shaded part corresponds to an $\{x, y\}$-split component which is neither a ring nor a bond. In each case, one can easily verify that the outer cycle satisfies Condition II(b).
Q.E.D.

Thus we concentrate on a graph having two or more critical separation pairs.

Lemma 4.5. *Let G be a 2-connected planar graph, and let $\{x, y\}$ be a prime separation pair of G. If a facial cycle S of G contains the vertices x and y, then exactly two $\{x, y\}$-split components contain edges of S.*

Proof. Since S is a cycle, at most two $\{x, y\}$-split components contains edges of S. Furthermore, since S is a facial cycle, not all the edges of S are contained in a single $\{x, y\}$-split component. Thus exactly two $\{x, y\}$-split components contain edges of S.
Q.E.D.

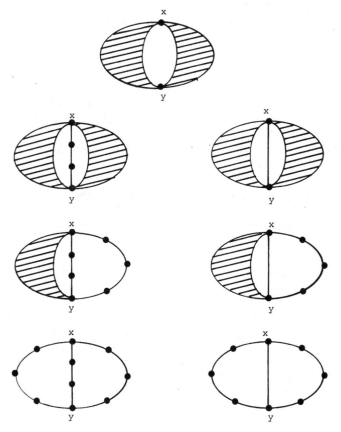

Fig. 4.5. Seven types of a plane graph having exactly one critical separation pair $\{x, y\}$.

We are now ready to present Theorem 4.3 which plays a crucial role in the testing algorithm.

Theorem 4.3. *Suppose that a 2-connected planar graph G has no forbidden separation pair and has two or more critical separation pairs. Apply the following operation to every critical separation pair $\{x, y\}$ of G: if $(x, y) \in E$, then delete edge (x, y) from G; if $(x, y) \notin E$ and exactly one $\{x, y\}$-split component is a ring, then delete the x–y path in the component from G. Let G_1 be the resulting graph. (Graphs G, and G_1 are illustrated in Figs. 4.6(a), and* (*b*), *respectively*.) *Then S is an extendible facial cycle of G if and only if S is a facial cycle of* G_1 *which contains all the vertices of critical separation pairs of G.*

Proof. *Necessity*: Assume that S is the extendible outer cycle of a *plane* graph G. Then, since S satisfies Condition II, all the deleted edges or paths are not on S, and hence S remains to be the outer cycle of the plane subgraph G_1 of G. Moreover S contains all the vertices of the critical separation pairs by Corollary 4.5.

Sufficiency: Assume that S is a facial cycle of G_1 which contains all the vertices of critical separation pairs of G. Clearly S is also a facial cycle of a plane embedding of G. We shall show that every critical separation pair $\{x, y\}$ of G satisfies Condition II(b). By Lemma 4.5, exactly two $\{x, y\}$-split components, say H_1 and H_2, contain edges of S. Therefore G has at most one $\{x, y\}$-split component containing no edges of S. Suppose that there exists such a component H_3 and that H_3 is neither a ring nor a bond. Then H_3 contains no vertex of critical separation pairs except x and y: if H_3 contains a vertex $z(\neq x, y)$ of a critical separation pair, H_3 would contain an edge of S since S contains all the vertices of critical separation pairs, contrary to the supposition. Therefore there is a critical separation pair $\{u, v\}$, different from $\{x, y\}$, such that vertex u or v is not contained in H_3. Note that G has at least two critical separation pairs. Hence H_1 or H_2, say H_1, is neither a ring nor a bond, and the other H_2 is a ring or a bond. Then edge (x, y) or the x–y path in H_2 should have been deleted in G_1, so H_2 could not contain edges of S, contrary to the assumption. Hence H_3 must be either a ring or a bond. Thus we have shown that S with G satisfies Condition II(b).   Q.E.D.

We define one more term before presenting two corollaries which follow immediately from Theorem 4.3. Let v be a vertex of a 2-connected plane graph G_2 and let $G_1 = G_2 - v$ be 2-connected. Then the v-*cycle* of G_2 is the cycle of the plane subgraph G_1 of G_2 which bounds the face of G_1 in which v lay.

Corollary 4.6. *Suppose that a 2-connected planar graph G has no forbidden separation pair and has two or more critical separation pairs. Let* G_1 *be the graph defined in Theorem 4.3. Let* G_2 *be the graph obtained from* G_1 *by adding a new vertex v and joining v to all the vertices of critical separation pairs of G. (Graph* G_2 *is illustrated in Fig. 4.6(c).) Then S is an extendible facial cycle of G if and only if*

(a) G_2 is planar; *and*

(b) S is the v-cycle of a plane embedding of G_2.

Corollary 4.7. *Suppose that a 2-connected planar graph G has no forbidden separation pair and has two or more critical separation pairs. Let G_2 be the graph defined in Corollary 4.6. Then G has a convex drawing if and only if G_2 is planar.*

Combining Corollaries 4.1, 4.2, 4.6, 4.7, and Lemma 4.4, we immediately have the following linear testing algorithm.

Algorithm CONVEX-TEST

Step 1. Find all the separation pairs of the given 2-connected planar graph G by the linear algorithm of Hopcroft and Tarjan [HT73] for finding 3-connected components. Determine three sets of separation pairs: the sets PSP of prime separation pairs; the set CSP of critical separation pairs; and the set FSP of forbidden separation pairs. If FSP $\neq \varnothing$ then stop with a message "G has no convex drawing" (see Corollary 4.1). If FSP = CSP = $\varnothing$ then print a message "All facial cycles are extendible" (see Corollary 4.2), and go to Step 3. If $|\text{CSP}| = 1$ then let S be an extendible outer facial cycle depicted in Fig. 4.5, and go to Step 3 (see Lemma 4.4).

Step 2. Obtain a graph G_1 from G by applying the operation in Theorem 4.3 to every critical separation pair of G. Construct a graph G_2 from G_1 by adding a new vertex v and joining v to all the vertices of CSP. Test the planarity of G_2 (see Corollary 4.7). If G_2 is nonplanar, then stop with a message "G has no convex drawing (There are no facial cycles containing every vertex of CSP)." Otherwise let S be the v-cycle of a plane graph G_2. Necessarily S satisfies Condition II (see Corollary 4.6).

Step 3. Stop with a message "The given graph G has a convex drawing."

Algorithm CONVEX-TEST finds an extendible facial cycle S if any. A

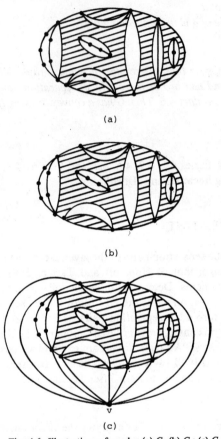

Fig. 4.6. Illustrations of graphs: (a) G; (b) G_1; (c) G_2.

method to generate all the extendible facial cycles of G is given in [CYN84]. A linear algorithm for drawing any planar graph "nicely" is given in [CON85].

4.4. Example

We illustrate a computational example in Fig. 4.7. Fig. 4.7(a) depicts an input graph G having 50 vertices and 83 edges. Algorithm CONVEX-TEST finds an extendible facial cycle C of G:

$$C = 15-16-1-2-3-4-5-6-7-8-9-10-11-12-13-14.$$

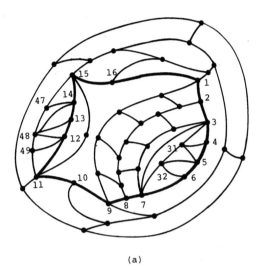

(a)

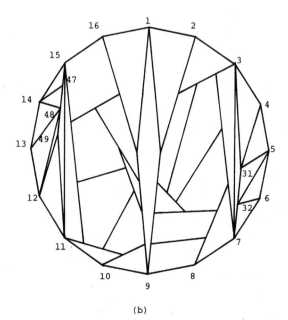

(b)

Fig. 4.7. A computational example: (a) input graph *G* (some vertex numbers are not shown for simplicity); and (b) output of CONVEX-DRAW.

The cycle C is indicated by bold lines in Fig. 4.7(a). When both a plane graph G with the outer cycle C and a convex polygon (regular 16-gon) of C are given, algorithm CONVEX-DRAW obtains a convex drawing of G, which is depicted in Fig. 4.7(b).

CHAPTER 5

VERTEX-COLORING

5.1. Introduction

A *(vertex-)coloring* of a graph is an assignment of colors to the vertices so that adjacent vertices have distinct colors. A *k-coloring* of a graph uses at most *k* colors. As an illustration we depict a 4-coloring of a graph in Fig. 5.1(a), where positive integers designate colors. That particular coloring is not optimal in the sense that there is another coloring using fewer colors; a 3-coloring is depicted in Fig. 5.1(b). The smallest integer *k* such that a graph *G* has a *k*-coloring is called the *chromatic number* of *G* and is denoted by $\chi(G)$. Thus the graph *G* in Fig. 5.1 satisfies $\chi(G) = 3$.

Fig. 5.1. (a) 4-coloring, and (b) 3-coloring.

The *graph coloring problem*, i.e., coloring a graph *G* with $\chi(G)$ colors, has practical applications in production scheduling, construction of time tables, etc. [Car79]. Since the problem is *NP*-hard [GJ79], it is unlikely that there exists any polynomial algorithm for the problem. One might expect that there would be an efficient approximate algorithm which uses a number of colors, not necessarily $\chi(G)$ but close to $\chi(G)$. However a polynomial algorithm that guarantees to color a graph with at most $a\chi(G) + b$ colors, $a < 2$, will imply a polynomial algorithm to color every graph *G* with $\chi(G)$ colors [GJ76]. In other words, getting close within a factor of two to the optimum is as hard as achieving it. The best known approximate coloring algorithm for a general graph may be found in [Wig83].

The situation for planar graphs is much more favorable. The famous four-color theorem recently proved by Appel and Haken says that every planar graph is 4-colorable. Since we do not supply the details of their proof, the interested reader should be referred to an original paper of Appel and Haken [AH77] and to the book of Saaty and Kainen [SK77]. We sketch only the outline of their proof. A graph is *k-chromatic* if it is not $(k-1)$-colorable but k-colorable. A *configuration* is an induced subgraph of a plane graph. A configuration is *reducible* if no minimal 5-chromatic plane graph can contain it. A set of configurations is *unavoidable* if every plane graph contains at least one configuration belonging to the set. In order to prove that every plane graph is 4-colorable, one sets out to find an *unavoidable set of reducible configurations*. Making use of the so-called discharging method and fast electronic computers, they eventually found an unavoidable set of over 1900 reducible configurations.

The proof of the four-color theorem leads to an algorithm of 4-coloring a plane graph. The algorithm runs in $O(n)$ recursive steps; at each step the algorithm detects in a graph one of over 1900 reducible configurations belonging to the unavoidable set, and recurses to a smaller graph. Since all the configurations contain at most 13 vertices, one recursive step can be done in time proportional to n, but the coefficient is no less than 1900. Thus the 4-coloring algorithm runs in $O(n^2)$ time, although it does not seem practical. The problem of finding a linear 4-coloring algorithm remains open.

In contrast, as we will see in Section 5.2, one can easily prove that every planar graph is 5-colorable, and there are very efficient algorithms which color any plane graph with at most five colors in linear time. We present two such algorithms in Sections 5.3 and 5.4.

Before ending this introduction, we add one more remark on the graph coloring problem. The problem remains *NP*-hard even for planar graphs [GJS76]. Every planar graph is 4-colorable, and it is easy to check whether a graph is 2-colorable, i.e., bipartite or not. Thus the problem of deciding whether a given planar graph is 3-chromatic or 4-chromatic is indeed *NP*-complete.

5.2. Proof of the five-color theorem and the $O(n^2)$ algorithm

In this section we first prove the five-color theorem, and then give an $O(n^2)$ algorithm of coloring any planar graph with at most five colors.

Although most of the standard texts on graph theory use the Kempe-chain argument in proving the theorem [Ber76, Bol79, BM76, CL85, Har72], we use an argument on "identification of vertices". An *identification* $\langle x, y \rangle$ of vertices x and y in a graph G means an operation on G which removes x and y and adds a new vertex adjacent to those vertices to which x and y was adjacent. A contraction of an edge (x, y) is a special case of the identification $\langle x, y \rangle$. For example a graph drawn by solid lines in Fig. 5.2(a) is transformed by identification $\langle v_6, v_2 \rangle$ into the one in Fig. 5.2(c).

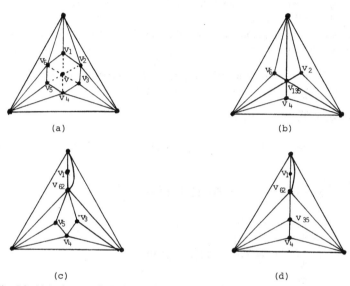

(a)

(b)

(c)

(d)

Fig. 5.2. (a) A plane graph G with a vertex v of degree 6, drawn by solid and dotted lines; (b) identification of v_1, v_3 and v_5 in $G - v$; (c) identification $\langle v_6, v_2 \rangle$ in $G - v$; (d) identification $\langle v_6, v_2 \rangle$ and $\langle v_3, v_5 \rangle$.

Theorem 5.1. *Every planar graph is 5-colorable.*

Proof. We prove the theorem by induction on the number n of vertices. The claim is trivially true for planar graphs with at most five vertices. Thus suppose that G is a planar graph with n vertices, and that all planar graphs with less than n vertices are 5-colorable. By Corollary 1.4 G contains a vertex v of degree at most five. There are two cases to consider.

Case 1: $d(v) \leqslant 4$. The deletion of v leaves us with a graph $G - v$ having $n - 1$ vertices which is 5-colorable by the inductive hypothesis. Then v can be colored with any color not used by the (at most four) neighbours, completing the proof in this case.

Case 2: $d(v) = 5$. Since G is planar, Kuratowski's theorem (Theorem 1.3) implies that the subgraph of G induced by the neighbours of v is not K_5, and hence v has two nonadjacent neighbours x and y. Delete vertex v from G, identify vertices x and y, and let G' be the resulting graph. Since G is planar, so is G'. Furthermore no loop is produced in G' since x and y are nonadjacent in G. Since G' has $n - 2$ vertices, by hypothesis G' has a 5-coloring, which naturally induces a 5-coloring of $G - v$ in which x and y are colored with the same color as the vertex substituting for x and y in G'. Assign to v any color other than the (at most four) colors of the neighbours, then we get a 5-coloring of G, completing the proof.   Q.E.D.

The proof above immediately yields the following recursive algorithm which colors any plane graph G with at most five colors.

Procedure FIVE-COLOR(G):
begin
   if $|V| \leqslant 5$ *then* find a trivial 5-coloring of G
   else
     begin
       if there is a vertex v of degree 4 or less
       then
         begin
           delete v from G;
           let G' be the resulting plane graph
         end
       else
         begin
           let v be a vertex of degree 5;
           let x and y be two nonadjacent neighbours of v;
           delete v from G, and identify x and y;
           let G' be the resulting graph
         end;
       FIVE-COLOR(G');
       assign to each identified vertex of G the color of the vertex substituting
       for it in G', and assign to v any color other than the (at most four) colors
       of the neighbours to complete a 5-coloring of G
     end
end.

Clearly all the operations used by the algorithm, other than deletions or identifications of vertices, require at most $O(n)$ time in total. If adjacency lists are used to represent a graph, the deletion of a single vertex v requires $O(d(v))$ time. Since $\sum_{v \in V} d(v) \leqslant 6n$ by Corollary 1.1, all deletions used in the algorithm spend at most $O(n)$ time in total. Thus the time required by identifications dominates the running time of the algorithm. One can easily execute a single identification $\langle x, y \rangle$ in $O(d(x) + d(y))$ time, that is, one can merge the two adjacency lists of x and y into a single list in that amount of time. However the same vertex may appear in identifications $O(n)$ times, so a direct implementation of the algorithm would require $O(n^2)$ time. In Chapter 6, we give a simple "on-line" algorithm to execute any sequence of identifications of a general graph G in $O(m \log n)$ time, by using adjacency lists together with an adjacency matrix. It immediately yields a simple $O(n \log n)$ 5-coloring algorithm.

In the succeeding two sections we discuss two linear algorithms. The first one given by Chiba, Nishizeki and Saito [CNS81b] runs in several stages, in each of which a set of identifications are performed in linear time but at least some fixed percentage of the vertices are eliminated. The second one given by Matula, Shiloach and Tarjan [MST80] and later simplified by Frederickson [Fre84] is to recurse after each identification, choosing identification that requires constant time to perform. Both approaches involve the clever exploitation of properties of planar graphs, and are good paradigms illustrating different techniques to speed up a planar algorithm.

5.3. Batch processing algorithm

The linear algorithm of [CNS81b] runs in several stages. Each stage processes a batch of vertices provided any vertex is not involved in more than two of identifications, so the stage requires $O(n)$ time. Furthermore the algorithm processes not only vertices of degree at most 5 but also those of degree 6 so that each stage eliminates at least $\frac{3}{17}$ of the vertices.

When G contains a vertex v of degree at most 4, construction of the reduced graph G' is the same as in FIVE-COLOR: simply $G' = G - v$. Otherwise the algorithm finds either
(a) a vertex of degree 5 such that at most one of the neighbours has been involved in identifications so far during the current stage, or
(b) a vertex v of degree 6 such that no neighbour of v has been involved in identifications so far during the current stage.
The following lemma implies that in Case (a) above vertex v necessarily has a pair of nonadjacent neighbours which have not been involved in identifications. In this case G' is constructed from $G - v$ by identifying the pair.

Lemma 5.1. *Let a planar graph $G = (V, E)$ contain a vertex v of degree 5 with $N(v) = \{v_1, v_2, v_3, v_4, v_5\}$. Then, for any specified $v_i \in N(v)$, v has a pair of nonadjacent neighbours v_j and v_k other than v_i. Furthermore one can find such a pair in $O(\min_{v \in N(v) - v_i} d(v))$ time if the planar embedding of G is given.*

Proof. We can assume without loss of generality that $v_i = v_1$, and that the vertices v_1, v_2, v_3, v_4 and v_5 in $N(v)$ are labelled clockwise about v in the plane embedding of G. Consider the case in which $d(v_2)$ is minimum among $d(v_2)$, $d(v_3)$, $d(v_4)$ and $d(v_5)$. Scanning the adjacency list of v_2, one can know whether $(v_2, v_4) \in E$ or not. If $(v_2, v_4) \in E$, then $(v_3, v_5) \notin E$. Thus one can find a pair of nonadjacent vertices in $O(d(v_2))$ time. The proof for all the remaining cases is similar to the above.   Q.E.D.

When a vertex v of degree 6 is found (i.e., Case (b) above occurs), the algorithm constructs G' from $G - v$ by identifying either three pairwise nonadjacent neighbours or two pairs of nonadjacent neighbours. Lemma 5.2 below guarantees that v has such neighbours that the resulting graph G' remains planar. Because either three of the six neighbours are identified into a single vertex or two pairs of neighbours are identified into two vertices, G' contains only *four* vertices of $N(v)$. Thus one can extend a 5-coloring of G' to a 5-coloring of G simply by assigning to v any color other than the (at most four) colors of the neighbours. As an illustration we depict in Fig. 5.2(a) a planar graph with a vertex v of degree 6. The vertex v has three pairwise nonadjacent neighbours v_1, v_3, v_5 and also two pairs $\{v_6, v_2\}$, $\{v_3, v_5\}$ of nonadjacent neighbours. Identifying v_1, v_3, v_5 into a single vertex v_{135}, i.e., two successive identifications $\langle v_1, v_3 \rangle$ and $\langle v_3, v_5 \rangle$, transform $G - v$ into a planar graph G' depicted in Fig. 5.2(b), while two identifications $\langle v_6, v_2 \rangle$ and $\langle v_3, v_5 \rangle$ transform $G - v$ into G' depicted in Fig. 5.2(d). Clearly identification of three neighbours into a single vertex necessarily transforms $G - v$ into a planar graph G'. However this is not the case when two pairs of nonadjacent neighbours are identified. For example $\langle v_6, v_2 \rangle$ and $\langle v_1, v_3 \rangle$ transform $G - v$ into a nonplanar graph G'. Thus the identified two pairs $\{v_i, v_j\}$ and $\{v_k, v_l\}$ must be "*parallel*", that is, v_i, v_j, v_k, v_l must appear in this order clockwise around v in the embedding. In the example above the two pairs $\{v_6, v_2\}$ and $\{v_3, v_5\}$ are parallel, while the two pairs $\{v_6, v_2\}$ and $\{v_1, v_3\}$ are not. Clearly identifications of two parallel pairs necessarily transform $G - v$ into a planar graph G'.

Lemma 5.2. *Let a planar graph $G = (V, E)$ contain a vertex v of degree 6 with $N(v) = \{v_1, v_2, \ldots, v_6\}$. Then v has either*
(i) *three pairwise nonadjacent neighbours,*
or

(ii) *two parallel pairs $\langle v_i, v_j \rangle$ and $\langle v_k, v_l \rangle$ of nonadjacent neighbours.*
Furthermore one can find these neighbours in $O(\min_{1 \leqslant s < t \leqslant 6}[d(v_s) + d(v_t)])$
time if the planar embedding of G is given.

Proof. Let $v_1, v_2, \ldots, v_6$ be the neighbours of v listed in cyclic order according
to the plane embedding of G. We establish our claim only for the case in which
$d(v_1) + d(v_2)$ is minimum among all the sums of degrees of two neighbours,
since the proof for all the remaining cases is similar. Scanning the adjacency
lists of v_1 and v_2, one can find whether the edges (v_1, v_5) and (v_2, v_4) exist or not.
If exactly one of them, say (v_1, v_5), exists, then v_2, v_4 and v_6 are the required three
pairwise nonadjacent neighbours. Otherwise, $\{v_6, v_2\}$ and $\{v_3, v_5\}$ (if both
(v_1, v_5) and (v_2, v_4) exist) or $\{v_2, v_4\}$ and $\{v_5, v_1\}$ (if neither exists) are the
required two "parallel" pairs of nonadjacent neighbours. Thus one can find the
required neighbours in $O(d(v_1) + d(v_2))$ time.   Q.E.D.

As a data structure to represent a graph G, we use an adjacency list Adj(v) for
each $v \in V$. Each adjacency list is doubly linked. The two copies of each edge
(u, v), one in Adj(u) and the other in Adj(v), are also doubly linked. In
addition to Adj, we use four arrays FLAG, COUNT, DEG and POINT
together with three queues $Q(i)$, $4 \leqslant i \leqslant 6$. An element DEG($v$) of array DEG
contains the value of $d(v)$, $v \in V$. FLAG(v) has an initial value "false" at the
beginning of each stage of the algorithm, and will be set to "true" when v is
identified with another vertex. COUNT(v) contains the number of neighbours
w with FLAG(w) = true, that is, the number of neighbours involved in
identifications in the current stage so far. The queue $Q(i)$, $4 \leqslant i \leqslant 6$, contains
all the vertices which are available for the recursive reduction of the stage,
defined as follows;

$$Q(4) = \{v \mid \text{DEG}(v) \leqslant 4\};$$

$$Q(5) = \{v \mid \text{DEG}(v) = 5, \text{COUNT}(v) \leqslant 1\}; \quad \text{and}$$

$$Q(6) = \{v \mid \text{DEG}(v) = 6, \text{COUNT}(v) = 0\}.$$

That is, $Q(4)$ is the set of all the vertices of degree 4 or less, $Q(5)$ the set of all
the vertices of degree 5 with at most one neighbour involved in an identification in the stage, and $Q(6)$ the set of all the vertices of degree 6 with no
neighbours involved in any identification in the stage. POINT(v) has a pointer
to an element "v" in $Q(i)$ if v is contained in $Q(i)$. We are now ready to present
the algorithm FIVE.

procedure FIVE;

```
begin
  embed a given planar graph G in the plane;
  for each v ∈ V do
    begin
      calculate DEG(v);
      FLAG(v) : = false;
      COUNT(v) : = 0
    end;
  initialize Q(4), Q(5) and Q(6);
  COLOR(G)
end.
```

The procedure COLOR called in FIVE runs in several stages, in each of which a batch of vertices are processed provided $Q(4)$, $Q(5)$, or $Q(6)$ is not empty.

```
procedure COLOR(G);
  {The procedures DELETE and IDENTIFY are for deletion and identifica-
  tion of vertices, respectively}
  begin
    if |V| ≤ 5 then find a trivial 5-coloring of G
    else
      begin
        if all Q(4), Q(5) and Q(6) are empty
          then {Current stage is over}
            reset FLAG, COUNT and Q; {Start a new stage}
        if Q(4) ≠ ∅
          then begin
            take a top entry v from Q(4);
            DELETE(v);
            let G' be the reduced graph
          end
          else
            if Q(5) ≠ ∅
              then begin
                take a top entry v from Q(5);
                choose two nonadjacent neighbours x, y such that FLAG(x)
                  = FLAG(y) = false;
                DELETE(v);
                IDENTIFY(x, y);
                let G' be the reduced graph
              end
```

else
  if $Q(6) \neq \varnothing$
    then
      begin
        take a top entry v from $Q(6)$;
        {By Lemma 5.2 either case (i) or case (ii) holds}
        for case (i) *do*
          begin
            let x, y and z be three pairwise nonadjacent
            neighbours of v;
            DELETE(v);
            IDENTIFY(y, x);
            IDENTIFY(z, x)
          end;
        for case (ii) *do*
          begin
            let $\{v_i, v_j\}$ and $\{v_k, v_l\}$ be two parallel pairs of
            nonadjacent neighbours;
            DELETE(v);
            IDENTIFY(v_i, v_j);
            IDENTIFY(v_k, v_l)
          end;
        let G' be the reduced graph
      end;

COLOR(G');
assign to each identified vertex of G the color of the vertex substituting
for it in G', and assign to v any color other than the (at most four) colors
of the neighbours;
  end
end;

procedure DELETE(v);
  begin
    for each $w \in \text{Adj}(v)$ *do*
    begin
      delete w from Adj(v);
      delete v from Adj(w);
      DEG(w) := DEG(w) $- 1$;
      if FLAG(v) = true
        then COUNT(w) := COUNT(w) $- 1$;

>   *end*;
>   delete Adj(v) from the adjacency lists and "v" from $Q(i)$, $i = 4, 5$ or 6, if
>   any, and update appropriately the elements in $Q(i)$ according to the modi-
>   fications of DEG and COUNT above
> *end*;

procedure IDENTIFY(u, v);
  {This procedure executes the identification $\langle u, v \rangle$ of two nonadjacent
  vertices u and v such that either FLAG(u) or FLAG(v) is "false". We assume
  FLAG(u) = false without loss of generality. The vertex v will act as a new
  vertex substituting for u and old v}
begin
  if FLAG(v) = false
    then begin
          FLAG(v) := true;
          for each $w \in$ Adj(v) *do* COUNT(w) := COUNT(w) + 1
        end;
    for each $w \in$ Adj(v) *do* mark w with "v";
    for each $w \in$ Adj(u) *do*
      begin
        delete w from Adj(u); delete u from Adj(w);
        if w has no mark "v"
        then begin
              {w is adjacent to u, but not to v}
              add w to Adj(v); add v to Adj(w);
              DEG(v) := DEG(v) + 1;
              COUNT(w) := COUNT(w) + 1;
              if FLAG(w) = true
              then COUNT(v) := COUNT(v) + 1
            end
        else {w is adjacent to both u and v}
              DEG(w) := DEG(w) − 1
      end;
    delete Adj(u) from the adjacency lists and "u" from $Q(i)$, $i = 4, 5$ or 6, if
    any, and update appropriately the elements in $Q(i)$, $i = 4, 5, 6$, according
    to the above modifications of DEG and COUNT
end;

In the algorithm above we omit the detail of the method for obtaining the
planar embedding of G' from that of G, since clearly the time required for the
purpose is proportional to that for deletions and identifications.

In order to show that algorithm FIVE runs in linear time we need the following lemma; the current version has been given by Frederickson [Fre84]. The lemma implies that at the end of each stage of the algorithm a positive fraction, say $\frac{1}{7}$, of the remaining vertices have been involved in identifications.

Lemma 5.3. *Let $G = (V, E)$ be a planar graph with minimum degree 5, and let S be a subset of V. If every vertex of degree 5 is adjacent to at least two vertices in S, and every vertex of degree 6 is adjacent to at least one vertex in S, then $|S| \geq (n + 12)/7$.*

Proof. Remember that n_k denotes the number of vertices having degree k. Let V_* be the set of vertices of degree greater than six, and n_* the number of such vertices. Let r_5 be the number of vertices of degree five in S, r_6 the number of vertices of degree six in S, and r_* the number of remaining vertices in S. From the conditions we have

$$\sum_{v \in S} d(v) \geq 2n_5 + n_6. \tag{5.1}$$

Note that

$$\sum_{v \in S} d(v) = 5r_5 + 6r_6 + \sum_{v \in S_*} d(v). \tag{5.2}$$

By definition we have

$$n_5 + n_6 + n_* = n$$

and from Corollary 1.1 we have

$$5n_5 + 6n_6 + \sum_{v \in V_*} d(v) \leq 6n - 12.$$

Subtracting the latter from seven times the former we have

$$2n_5 + n_6 \geq n + 12 + \sum_{v \in V_*} (d(v) - 7)$$

$$\geq n + 12 + \sum_{v \in S_*} (d(v) - 7)$$

$$= n + 12 + \sum_{v \in S_*} d(v) - 7r_*. \tag{5.3}$$

Combining (5.1), (5.2) and (5.3) yields

$$5r_5 + 6r_6 + \sum_{v \in S_*} d(v) \geq n + 12 + \sum_{v \in S_*} d(v) - 7r_*,$$

from which the claim immediately follows.   Q.E.D.

We are now ready to prove the following theorem.

Theorem 5.2. *The procedure FIVE colors a planar graph G with at most five colors in* $O(n)$ *time.*

Proof. Similarly as in the proof of Theorem 5.1, one can easily prove by induction on the number of vertices of a graph that the algorithm correctly colors a planar graph G with at most five colors. Hence we shall verify the claim on time complexity.

We first show that the first stage of the algorithm requires at most $O(n)$ time. One can easily verify that the procedure DELETE executes the deletion of a vertex v in $O(d(v))$ time, and that the procedure IDENTIFY does the identification of two nonadjacent vertices u and w in $O(d(u) + d(w))$ time since it simply scans the elements of Adj(u) and Adj(w). The algorithm calls DELETE for a vertex in each reduction. Since every vertex appears in at most one deletion, all the vertex-deletions in the stage require $O(n)$ time in total. Consider a reduction around a vertex v of degree 5 or 6, in which IDENTIFY is called in addition to DELETE. If v is in $Q(5)$, the algorithm finds two neighbours v_i and v_j of v with FLAG(v_i) = FLAG(v_j) = false, and then calls IDENTIFY(v_i, v_j). The identification requires $O(d(v_i) + d(v_j))$ time. Lemma 5.1 implies that one can find v_i and v_j in that amount of time. If v is in $Q(6)$, the algorithm finds either three pairwise nonadjacent neighbours x, y and z or two pairs of nonadjacent neighbours $\{v_i, v_j\}$ and $\{v_k, v_l\}$, and then calls IDENTIFY(y, x) and IDENTIFY(z, x) or IDENTIFY(v_i, v_j) and IDENTIFY(v_k, v_l), respectively. These two identifications together require $O(d(x) + d(y) + d(z))$ or $O(d(v_i) + d(v_j) + d(v_k) + d(v_l))$ time, respectively. Lemma 5.2 implies that one can find these neighbours in that amount of time. Of course, FLAG's for these vertices are all "false", since COUNT(v) = 0. That is, all these vertices have not been involved in any identification in the stage. Thus every vertex is involved in at most two identifications in the stage. (The vertex x above is involved in two identifications.) Therefore all the identifications in the stage require $O(n)$ time in total. Clearly the book-keeping operations required for the four arrays and three queues need $O(n)$ time in total. Note that one can directly access "v" via a pointer in POINT(v). Hence we can conclude that the stage requires $O(n)$ time.

We next show that at the end of the first stage the reduced graph $G' = (V', E')$ contains at most $14n/17$ vertices. Suppose that $|V'| = n' \neq 0$. Then the minimum degree of G' is 5, and COUNT(v) $\geqslant 2$ for every vertex v of degree 5, and COUNT(v) $\geqslant 1$ for every vertex of degree 6, since $Q(4)$, $Q(5)$

and $Q(6)$ are all empty at the end of the stage. Let $S = \{v \mid \text{FLAG}(v) = \text{true},\ v \in V'\}$, then the subset S of V' satisfies the requirement of Lemma 5.3, and hence $|S| \geqslant n'/7$. Clearly at least $|S|$ vertices disappear from the graph G by identifications. Since each reduction produces at most two vertices in S, there must occur at least $|S|/2$ graph reductions around vertices of degree 5 or 6 in the stage. Therefore at least $|S|/2$ vertices are removed from G by deletions in the stage. Hence at least $3|S|/2$ vertices disappear from G in the stage. Therefore we have

$$n - n' \geqslant 3|S|/2.$$

Since $|S| \geqslant n'/7$, we have

$$n' \leqslant 14n/17.$$

Using the two facts above, we have the following equations on $T(n)$ the time needed to 5-color a planar graph G of n vertices:

$$T(n) \leqslant c_1 \qquad \qquad \text{if } n \leqslant 5;$$

$$T(n) \leqslant T(14n/17) + c_2 n \quad \text{otherwise,}$$

where c_1 and c_2 are constants. Solving these equations, we have $T(n) = O(n)$.

Q.E.D.

5.4. Sequential processing algorithm

In this section we present another linear 5-coloring algorithm originally given by Matula, Schiloach and Tarjan [MST80] and later simplified by Frederickson [Fre84]. The algorithm is a rather straightforward improvement of the $O(n^2)$ algorithm in Section 5.2. That is, it sequentially processes only vertices of degree 5 or less. However it must find a vertex v of degree 5 having a pair of nonadjacent neighbours whose identification can be done in constant time. The following lemma guarantees the existence of such v. Remember that identification $\langle x, y \rangle$ can be performed in $O(d(x) + d(y))$ time.

Lemma 5.4. *Let G be a planar graph of minimum degree five. Then there exists a vertex v with $d(v) = 5$ having a pair of nonadjacent neighbours x, y such that $d(x), d(y) \leqslant 7$.*

Proof. Consider a maximal plane graph G', which is augmented from G by adding edges. A desired vertex in G' will also be a desired vertex in G. For each vertex v with $d(v) = 5$, let $v_0, v_1, \ldots, v_4$ be the neighbours of v listed in cyclic order according to the embedding of G'. Since G' is maximal planar, v_i and $v_{(i+1)\bmod 5}$ are adjacent for each i. There are two cases to consider.

Case 1: there is no vertex v of degree 5 such that v_i and $v_{(i+2)\bmod 5}$ are adjacent for some i. Assume that G' contains no desired vertex. Then every vertex v of degree 5 has at most two neighbours of degree $\leqslant 7$; furthermore if v has two, they must be successive in the cyclic order. Thus for each vertex v of degree 5 there must be a cyclic indexing of the v_j such that v_0, v_1, v_2 are of degree $\geqslant 8$. Let d_8 be the total degree of vertices having degree $\geqslant 8$, that is, $d_8 = \Sigma_{i \geqslant 8} in_i$. We now claim that each vertex v of degree 5 contributes to d_8 at least 5. The edges (v, v_0), (v, v_1), and (v, v_2) each contribute one to d_8, yielding a total of three. In addition, edges (v_0, v_1) and (v_1, v_2) each contribute two to d_8, or four altogether. But since each of these edges is shared with one other face, just half of the contribution can be apportioned to v, yielding a total of two from edges (v_0, v_1) and (v_1, v_2). Thus altogether v contributes at least 5 to d_8. Thus we have

$$5n_5 \leqslant d_8 = \sum_{i \geqslant 8} in_i.$$

However from Corollary 1.3 we have

$$n_5 \geqslant 12 + \sum_{i \geqslant 7} (i-6)n_i > \sum_{i \geqslant 8} in_i/4.$$

This contradicts the above derived inequality, so that a desired vertex must exist in this case.

Case 2: there is some vertex v of degree 5 such that there is a cyclic indexing of its neighbours in which v_2 and v_4 are adjacent. Call such a vertex v *special*. Clearly vv_2v_4 is a nonfacial triangle. One may assume that vv_2v_4 is innermost in a sense that no such triangle is contained in vv_2v_4. We show that there is a desired vertex in the triangle.

Let G_v be the subgraph of G induced by the vertices on and in the triangle. Clearly $d(G_v, v) = 3$, and $d(G_v, v_2)$, $d(G_v, v_4) \geqslant 3$ since v_2 and v_4 are adjacent to each other and to v_3. All the remaining vertices of G_v are of degree at least 5. Using G_v as a building block, we shall generate a graph of minimum degree five. Consider the octahedral graph depicted in Fig. 5.3. Construct G_8 by embedding a copy of G_v in each face of the octahedral graph, identifying vertices v, v_2, and v_4 and edges (v, v_2), (v, v_4), and (v_2, v_4) of each G_v with the vertices and edges bounding the face in which it is embedded. Each vertex of G_8 originally in the octahedral graph will have degree at least eight, since v, v_2, and v_4 are each of degree at least three in G_v.

Since there is no special vertex in G_v other than v, v_2 or v_4, and the octahedral

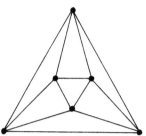

Fig. 5.3. Octahedral graph.

vertices are of degree at least eight, there are no special vertices in G_8. Since G_v is maximal planar, so is G_8. Thus G_8 satisfies the condition in Case 1 and thus contains a desired vertex. From the construction it corresponds to a desired vertex in G'.   Q.E.D.

Thus, when the minimum degree is five, the algorithm finds a vertex v of degree 5 with a pair of nonadjacent neighbours $\{x, y\}$ such that $d(x), d(y) \leqslant 7$. The identification $\langle x, y \rangle$ can be performed in constant time. Since at most n identifications occur, they spend $O(n)$ time in total. Using data structure similar to that in the batch processing algorithm of the previous section, one can easily implement the algorithm to run in linear time. The algorithm has a merit: it need not have any plane embedding of G.

As we have seen, two distinct properties of planar graphs mentioned in Lemmas 5.3 and 5.4 are crucial in the design of two linear 5-coloring algorithms.

CHAPTER 6

EDGE-COLORING

6.1. Introduction

In this chapter we discuss the edge-coloring problem for *graphs* and *multigraphs*. The problem is simply stated: color the edges of a given (multi)graph G using as few colors as possible, so that no two adjacent edges receive the same color. The problem arises in many applications, including various scheduling and partitioning problems [FW77]. In view of the potential applications, it would be useful to have an efficient algorithm capable of coloring any (multi)graph G with this minimum number of colors (called the *chromatic index* of G and denoted by $q^*(G)$). Unfortunately no such efficient algorithm is currently known. Moreover Holyer has shown that the edge-coloring problem is *NP*-hard [Hol81], and therefore it seems unlikely that any such polynomial-time algorithm exists. He actually showed that the problem is *NP*-hard even if restricted to the cubic graphs. Since any cubic graph can be edge-colored with three or four colors, his result implies that, unless $P = NP$, there does not exist a polynomial-time approximation algorithm which finds an edge-coloring with at most $(\frac{4}{3} - \varepsilon)q^*(G)$ colors for any $\varepsilon > 0$ (see Lemma 6.6).

Let $d(G)$ (or simply d) denote the maximum degree of a (multi)graph G. By a well-known result due to Vizing, the chromatic index of a (simple) graph is at most $d(G) + 1$ [FW77, Viz64]. Special cases which can be colored with d colors are bipartite graphs, cubic bridgeless planar graphs (whose edge-coloring in three colors is equivalent to the four color problem) and planar graphs with $d \geqslant 8$ [CH82, FW77, GK82]. The situation for multigraphs is more complicated. The best known upper bound for the chromatic index $q^*(G)$ of multigraphs G is given by Nishizeki and Kashiwagi [NK84]:

$$q^*(G) \leqslant \max\{p(G), \lfloor 1.1d(G) + 0.8 \rfloor\}.$$

Here $p(G)$ is defined as follows:

$$p(G) = \max_{H \subset G} \lceil m(H)/\lfloor n(H)/2 \rfloor \rceil,$$

where H runs over all subgraphs of G having three or more vertices, $m(H)$ is the number of edges in H, and $n(H)$ the number of vertices in H. Since at most $\lfloor n(H)/2 \rfloor$ edges of H can be colored with the same color, $p(G)$ is clearly a trivial lower bound on $q^*(G)$. The upper bound above follows a long history of research dating back to Shannon [Sha49], who proved the following upper bound: $q^* \leqslant \lfloor 3d/2 \rfloor$. His bound has been successively improved by Goldberg [Gol73], Andersen [And77], Goldberg [Gol84a] and Nishizeki and Kashiwagi [NK84].

In the succeeding two sections we present two algorithms for edge-coloring graphs [GNKLT84]. One of them, COLOR, edge-colors any graph G with d or $d + 1$ colors. The other ALCOLOR edge-colors an arbitrary graph G with d or $d + 1$ colors and edge-colors with d colors a large class of graphs including: (1) all the planar graps with $d \geqslant 8$; (2) all the series-parallel graphs with $d \geqslant 4$; and (3) almost all random graphs. Algorithm ALCOLOR employs COLOR together with an argument used in Vizing's "Adjacency Lemma" for critical graphs. Both algorithms run in $O(mn)$ time and use $O(m)$ space. As an approximation algorithm for graphs, COLOR is best possible in the sense that it uses at most one extra color. The fastest known algorithm for coloring a graph with $d + 1$ colors runs in $O(dm \log n)$ or $O(m(n \log n)^{1/2})$ time [Arj82, GNKLT84].

In Section 6.4 we present an approximation algorithm MULTICOLOR for multigraphs, given by Nishizeki and Sato [NS84]. The algorithm finds an edge-coloring of any given multigraph G, using "close" to the optimal number $q^*(G)$ of colors. More precisely MULTICOLOR uses at most $q(G)$ colors, where

$$q(G) = \max\{p(G), \lfloor (5d(G) + 2)/4 \rfloor\}.$$

It spends $O(m(d + n))$ time and $O(m)$ space. MULTICOLOR achieves the worst-case ratio $\max_G q(G)/q^*(G)$ of $\frac{4}{3}$, which looks to be the best possible as a polynomial-time approximation algorithm. Other algorithms using at most $\max\{p(G), \lfloor (9d(G) + 6)/8 \rfloor\}$ or $\max\{p(G), \lfloor 1.1d(G) + 0.8 \rfloor\}$ colors are known [Gol84b, HNS84, NK84], but they are too complicated to be included in this text.

6.2. Algorithm COLOR

In this section we present an algorithm COLOR for coloring a graph with d or $d + 1$ colors on the basis of a constructive proof of Vizing's theorem [Bol79,

TN82]. We now introduce some terminology. A graph obtained from G by adding edge (u, v) is denoted by $G + (u, v)$. An edge-coloring with at most k colors is called a *k-coloring*. Suppose that G is colored with a set of colors. If color c in the set is not used for any edge incident with vertex v, then c is *missing* at v or v *misses* color c. We denote by $M(v)$ the set of all the colors missing at v. Consider the subgraph of G induced by the edges colored with color c or c'. Clearly each component of the subgraph is a path or cycle, in which edges are colored with c or c' alternately. Such a path is called a *cc'-alternating path*. A vertex $x \in V$ is an end of such a path if and only if c or $c' \in M(x)$. If $c \in M(x)$ for a vertex $x \in V$, then Apath(x, c, c') denotes a cc'-alternating path starting from x. Interchanging colors c and c' in Apath(x, c, c') yields another coloring of G with the same set of colors, in which x misses color c'.

Suppose that all the edges of G except (v, w) have been colored with a set of $d + 1$ colors. Clearly both v and w have at least two missing colors, and each of the others has at least one. We associate with each vertex x *one* of the missing colors of x, denoted by mis(x) and called *the* missing color of x. A *fan sequence* F at w starting with (w, v) is a sequence of distinct edges $(w, x_0) = (w, v), (w, x_1), \ldots, (w, x_s)$ such that for each $1 \leqslant i \leqslant s$ wx_i is colored with mis(x_{i-1}). (See Fig. 6.1.) If F is a maximal fan sequence at w, then one of the following must occur:

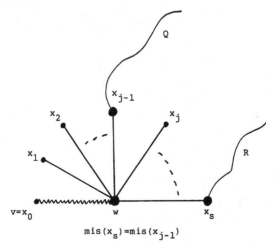

mis(x_s) = mis(x_{j-1})

Fig. 6.1. A fan sequence and paths Q and R.

Case (1): the missing color of x_s is also missing at w, that is, mis$(x_s) \in M(w)$; or

Case (2): an edge of F is colored with $\text{mis}(x_s)$.

Note that F consists of a single uncolored edge (w, v) if $\text{mis}(v) \in M(w)$. We are now ready to present COLOR.

procedure COLOR(G):
begin
  $G' := (V, \varnothing)$ {empty graph}
  for each edge $e = (v, w)$ of G *do*
    begin
      $G' := G' + e$
      RECOLOR(G', e); {update a coloring of $G' - e$ to G'}
    end
end;

The procedure RECOLOR extends a $(d + 1)$-coloring of $G' - e$ into a $(d + 1)$-coloring of G'.

procedure RECOLOR(G', e):
begin
  let $F = [(w, x_0) = (w, v), (w, x_1), \ldots, (w, x_s)]$ be a maximal fan sequence F at w starting with the uncolored edge (w, v);
  if $\text{mis}(x_s) \in M(w)$ {Case (1)} *then* {shift colors of F}
    for $i := 0$ *to* s *do* color (w, x_i) with $\text{mis}(x_i)$
  else {Case (2)}
    begin
      let (w, x_j), $1 \leqslant j \leqslant s - 1$, be the edge colored with $\text{mis}(x_s)$;
      {$\text{mis}(x_{j-1}) = \text{mis}(x_s)$}
      if path $Q = \text{Apath}(x_{j-1}, \text{mis}(x_s), \text{mis}(w))$ does not end at w
      then
        begin
          interchange the colors of Q;
          {color $\text{mis}(w)$ is now missing both at w and x_{j-1}}
          for $i := 0$ *to* $j - 2$ *do*
            color (w, x_i) with $\text{mis}(x_i)$;
          color (w, x_{j-1}) with $\text{mis}(w)$
        end
      else {since Q ends at w, path $R = \text{Apath}(x_s, \text{mis}(x_s), \text{mis}(w))$ does not end at w}
        begin
          interchange the colors of R;

{mis(w) is now missing both at w and x_s}
for $i := 0$ *to* $s - 1$ *do*
    color (w, x_i) with mis(x_i);
    color (w, x_s) with mis(w)
    end
  end
end;

We have the following theorem on COLOR.

Theorem 6.1. *Algorithm COLOR edge-colors an arbitrary graph* $G = (V, E)$ *with* d *or* $d + 1$ *colors in* O(nm) *time, using* O(m) *space.*

Proof. (a) *Correctness*: It is rather easy to see the fact that procedure RECO-LOR extends a $(d + 1)$-coloring of $G' - e$ into a $(d + 1)$-coloring of G'. Using this fact, one can immediately prove the correctness of the algorithm by the induction on m.

(b) *Space*: We use the following data structures.

(*1*) *Adjacency lists*: G is represented by adjacency lists, each containing the edges incident with a vertex. These use O(m) space.

(*2*) *An array of missing colors*: The function mis($\cdot$) is represented by an array mis[$\cdot$] of length n.

(*3*) *Color lists*: Each list contains the edges colored with the same color. Thus these $d + 1$ lists use O(m) space.

An edge (x, y) colored with c appears in the adjacency lists for x or y and also in the color list for c. These three elements are linked to each other by pointers so that each can be directly accessed from another. Clearly these devices use O(m) space in total.

(*c*) *Time*: It suffices to show that one execution of RECOLOR can be done in O(n) time, since COLOR repeats RECOLOR m times.

We first show that the fan sequences F at w can be found in O(d) time. Construct an array W of length $d + 1$ such that $W[c]$ contains either the edge incident with w and colored with c or a mark indicating that $c \in M(w)$. Exploring the adjacency list for w, we can construct W in O(d) time. Then we can decide in O(1) time whether a given color c is missing at w, and also find the edge colored with c if $c \notin M(w)$. Using arrays mis[$\cdot$] and $W[\cdot]$, we can easily find F in O(d) time.

Using the color lists, we can easily construct the subgraph induced by edges colored mis(x_s) or mis(w) in O(n) time, since it contains at most n edges. Furthermore, using the depth-first search (in Chapter 2), we can decide in O(n) time whether Apath(x_{j-1}, mis(x_s), mis(w)) ends at w.

Clearly we can initialize mis[·] and the color lists for an empty graph in $O(n)$ time. Furthermore we can easily update mis[·] and the color lists for $G' - e$ into ones for G' in $O(n)$ time.

Thus we have shown that one execution of RECOLOR can be done in $O(n)$ time.

Q.E.D.

6.3. Algorithm ALCOLOR

In this section we give algorithm ALCOLOR which edge-colors an arbitrary graph with d or $d + 1$ colors and with d colors for a large class of graphs.

ALCOLOR is based on the proof of Vizing's "Adjacency Lemma" [FW77, Viz65a, Viz65b]. Denote by $d^*(w)$ the number of vertices adjacent to w and having degree d. Then an edge (v, w) is defined to be *eliminable* if w has at most $d - d(v)$ neighbours of degree d other than v, that is,

$$d(v) + d^*(w) \leqslant d \quad \text{when } d(v) < d; \text{ and}$$

$$d^*(w) = 1 \qquad \qquad \text{when } d(v) = d.$$

In other words, the edges that are excluded in a "critical" graph by the adjacency lemma are eliminable. (Notice that the definition is not symmetric with v and w.)

ALCOLOR is outlined as follows. It repeatedly deletes eliminable edges from G until they all disappear or the maximum degree decreases. Let G' be the resulting graph. An edge that was not eliminable in original G may become eliminable when some edges have deleted. On the other hand, once an edge becomes eliminable, it remains so thereafter. There are two cases: $d(G') = d(G) - 1$; or $d(G') = d(G)$. In the lucky case $d(G') = d(G) - 1$, we first color G' with $d(= d(G') + 1)$ colors by algorithm COLOR. We then update the d-coloring of G' to a d-coloring of G' plus the last deleted edge, using procedure ALCOLOR. Repeat this updating for each of the deleted edges in the *reverse* order until a d-coloring of G is obtained. In the unlucky case when $d(G') = d(G)$, we simply color G itself with $d + 1$ colors by COLOR.

procedure ALCOLOR(G):
begin
  $d := d(G)$;
  $G' := G$;
  while G' has an eliminable edge (v, w) *and* $d(G') := d$ *do*
    begin
      $G' := G' - (v, w)$; {delete (v, w)}

push (v, w) on the top of stack S;
   end;
if $d(G') = d - 1$ {lucky case} *then*
   begin
      COLOR(G'); {G' is now colored with $d = d(G') + 1$ colors}
      while stack S is not empty *do*
         begin
            pop up an edge, say (v, w), from S;
            $G' := G' + (v, w)$;
            ALRECOLOR$(G', (v, w))$ {update d-coloring}
         end
   end
   else {unlucky case: G' has no eliminable edge and $d(G') = d$}
      COLOR(G)
end;

We now explain procedure ALRECOLOR. Suppose that edge (v, w) is eliminable in G and that $G - (v, w)$ is colored with a set of d colors. In the d-coloring of $G - (v, w)$, vertex v has $d - d(v) + 1 (\geqslant 1)$ missing colors; vertex $u (\neq v, w)$ has at least one if $d(u) < d$ and has no missing color if $d(u) = d$. We associate with each vertex $u \in V$ a missing color of u if any, which is denoted by $\text{mis}(u)$. For each c of the $d - d(v) + 1$ missing colors of v, there exists a maximal fan sequence $F = [(w, x_0) = (w, v), (w, x_1), \ldots, (w, x_s)]$ at w in which edge (w, x_1) is colored with c if $s \geqslant 1$. One of the following must occur:
(a) $\text{mis}(x_s) \in M(w)$;
(b) an edge of F is colored with $\text{mis}(x_s)$;
(c) $d(x_s) = d$ (and hence x_s has no missing color).
Since (v, w) is eliminable, w has at most $d - d(v)$ adjacent vertices of degree d other than v. Therefore one of the following must occur:

Case (1): there exists a maximal fan sequence

$$F = [(w, x_0), (w, x_1), \ldots, (w, x_s)]$$

such that $\text{mis}(x_s) \in M(w)$;

Case (2): there exists a maximal fan sequence

$$F = [(w, x_0), (w, x_1), \ldots, (w, x_s)]$$

such that an edge of F is colored with $\text{mis}(x_s)$;

Case (3): there exist two fan sequences $F_1 = [(w, x_0), (w, x_1), \ldots, (w, x_s)]$ and $F_2 = [(w, y_0), (w, y_1), \ldots, (w, y_t)]$ which meet exactly at $v(= x_0 = y_0)$ and $x_s(= y_t)$. (See Fig. 6.2. F_1 and F_2 are not always maximal fans.)

Because, if none of the three cases above occurred, then the $d - d(v) + 1$ maximal fan sequences were edge-disjoint except for the starting edge (v, w) and each of them ended with an edge having an end $(\neq w)$ of degree d, and hence w would have at least $d - d(v) + 1$ neighbours of degree d, a contradiction.

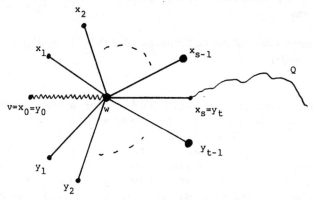

Fig. 6.2. Two intersecting fan sequences and path Q.

We are now ready to present ALRECOLOR.

procedure ALRECOLOR$(G, (v, w))$:
begin {the uncoloured edge (v, w) of G is eliminable}
    if Case (1) or (2) occurs *then*
        extend a d-colouring of $G - (v, w)$ into a d-colouring of G as in Case (1) or
        (2) of RECOLOR
    else {Case (3) occurs}
        begin
            {$\mathrm{mis}(x_{s-1}) = \mathrm{mis}(y_{t-1})$}
            if path $Q = $ Apath$(w, \mathrm{mis}(w), \mathrm{mis}(x_{s-1}))$ does not end at x_{s-1} *then*
            begin
                color (v, w) with the color of (x_1, w);
                for $i := 1$ *to* $s - 2$ *do*
                    color (w, x_i) with $\mathrm{mis}(x_i)$;
                erase the color of (w, x_{s-1});

{now (w, x_{s-1}) is uncolored}
interchange the colors of Q;
{now both w and x_{s-1} miss color $\text{mis}(x_{s-1})$}
color (w, x_{s-1}) with $\text{mis}(x_{s-1})$
end
else {Q does not end at y_{t-1}}
begin
color (v, w) with the color of (y_1, w);
for $i := 1$ *to* $t - 2$ *do*
color (w, y_i) with $\text{mis}(y_i)$;
erase the color of (w, y_{t-1});
{now (w, y_{t-1}) is uncolored}
interchange the colors of Q;
{now both w and y_{t-1} miss color $\text{mis}(y_{t-1}) = \text{mis}(x_{s-1})$}
color (w, y_{t-1}) with $\text{mis}(y_{t-1})$
end
end
end;

We immediately have the following lemma on ALRECOLOR.

Lemma 6.1. *Suppose that all the edges of G except (v, w) are colored with d colors and that (v, w) is eliminable in G. Then ALRECOLOR edge-colors G with d colors in $O(n)$ time, using $O(m)$ space.*

Then we have the following theorem.

Theorem 6.2. *Algorithm ALCOLOR edge-colors an arbitrary graph G with d or $d + 1$ colors in $O(nm)$ time, using $O(m)$ space.*

Proof. Since we can easily prove the correctness of ALCOLOR, we shall establish the claims on time and space. Additional space needed for ALCO-LOR is the stack S to store the deleted edges, which clearly uses $O(m)$ space. In order to find eliminable edges we need two arrays $d[\cdot]$ and $d^*[\cdot]$, each of length n, representing $d(u)$ or $d^*(u)$ for $u \in V$. Thus ALCOLOR uses $O(m)$ space.

One execution of ALRECOLOR can be done in $O(n)$ time, and ALRECO-LOR is executed at most m times. Therefore ALRECOLOR spends $O(nm)$ time in total. Since we can delete an edge in $O(1)$ time under our data structure, the deletion of edges spends $O(m)$ time in total. Thus we shall show that we can find eliminable edges in $O(nm)$ time in total. Since we can compute $d(u)$ and

$d^*(u)$ for all $u \in V$ in $O(m)$ time, we can find all the edges eliminable in original G in $O(m)$ time. Hence it suffices to show that the newly eliminable edges can be found in $O(nm)$ time in total.

Suppose that edge (x, y) is deleted in graph G'. Then we update $d[\cdot]$, that is, decrease both $d[x]$ and $d[y]$ by one. Some of the edges incident on x or y may become eliminable. Therefore we check for each of the edges incident on x or y whether it becomes eliminable. Clearly the checking can be done in $O(n)$ time. Since ALCOLOR deletes at most m edges, this checking can be done in $O(nm)$ time in total.

However the above procedure does not find all the edges that become newly eliminable. Suppose that $d(x)$ or $d(y)$, say $d(x)$, equals d. Then $d^*(z)$ decreases for each neighbour z of x, and so an edge incident with z may become eliminable. Therefore we must check for every edge incident on z whether it becomes eliminable. This checking can be done in $O(d(x) + \Sigma_{z \in N(x)} d(z)) = O(m)$ time. Of course the arrays $d[\cdot]$ and $d^*[\cdot]$ can be updated in this time. The case in which an end of a deleted edge has degree d occurs at most n times. Hence the checking above can be done in $O(nm)$ time in total.

Thus we have shown that the newly eliminable edges can be found in $O(nm)$ time in total.   Q.E.D.

In the remainder of this section we will discuss the features of ALCOLOR for planar graphs, series-parallel graphs and random graphs.

6.3.1. Planar graphs

Vizing [Viz65a] has proved that $q^*(G) = d$ if G is a planar graph with $d \geq 8$. There exists a planar graph G with $q^*(G) = d + 1$ for each d, $3 \leq d \leq 5$, while it is conjectured that $q^*(G) = d$ for every planar graph G of $d = 6$ or 7 [FW77]. Algorithm ALCOLOR colors any planar graph with d colors if $d \geq 8$. We have the following lemma.

Lemma 6.2. *Any planar graph of maximum degree $d \geq 8$ has an eliminable edge.*

Proof. Suppose that a planar graph G with $d \geq 8$ has no eliminable edges. Let n_i be the number of vertices of degree i in G. Clearly $n_1 = 0$. Since G is planar, by Corollary 1.3

$$12 + n_7 + 2n_8 + \cdots + (d - 6)n_d \leq 4n_2 + 3n_3 + 2n_4 + n_5. \quad (6.1)$$

Let $n_d(i_2, i_3, \ldots, i_7)$ be the number of vertices of degree d which have i_2

neighbours of degree 2, i_3 of degree 3, ... , i_7 of degree 7. Since each edge (v, w) of G is not eliminable,

$$d(v) + d^*(w) \geq d + 1 \quad \text{if } d(v) < d;$$

$$d^*(w) \geq 2 \qquad\qquad \text{if } d(v) = d.$$

Clearly $d^*(v) \geq 2$ for every $v \in V$. Assume that $2 \leq j \leq 7 (\leq d - 1)$. Counting the number of edges with one end of degree j and the other end of degree d, we have

$$2n_j \leq \sum i_j n_d(i_2, i_3, \ldots, i_7),\tag{6.2}$$

where the summation is over all possible $i_2, i_3, \ldots, i_7$.

Equation (6.2) can be refined if $j = 3$ or 4 in particular [Yap81]. First let v be any vertex of degree 3, and consider, in detail, the degrees of neighbours of v. For any neighbour w of v, we have $d^*(w) \geq d - 2$ and hence $d(w) = d$ or $d - 1$. Therefore the following must occur.

(a) One of the neighbours of v has degree $d - 1$ and the other two have degree d; or

(b) The three neighbours have degree d.

Let r be the number of vertices of degree 3 satisfying (a). Then (6.2) is refined for $j = 3$ as follows:

$$2r + 3(n_3 - r) \leq \sum i_3 n_d(i_2, i_3, \ldots, i_7).\tag{6.3}$$

Next let v be any vertex of degree 4. Let w_1, w_2, w_3, and w_4 be the neighbours of v. Since (v, w_i) is not eliminable, $d^*(w_i) \geq d - 3$ and hence $d(w_i) \geq d - 2$ for $i = 1, 2, 3, 4$. If $d(w_i) = d - 2$ for some i, then $d^*(v) \geq 3$, and hence $d(w_j) = d$ for all $j \neq i$. Therefore the following must occur.

(a) One of the neighbours of v has degree $d - 2$, and the other three have degree d;

(b) one has degree $d - 1$ and the other three have degree d;

(c) two have degree $d - 1$ and the other two have degree d; or

(d) all the four have degree d.

Let s, t, and u be the numbers of vertices of degree 4 satisfying (a), (b) and (c), respectively. Then (6.2) is refined for $j = 4$ as follows:

$$3s + 3t + 2u + 4(n_4 - s - t - u) \leq \sum i_4 n_d(i_2, i_3, \ldots, i_7).\tag{6.4}$$

Thus from (6.2), (6.3) and (6.4) we have

$$2n_2 + \{2r + 3(n_3 - r)\}/2 + \{3s + t + 2u + 4(n_4 - s - t - u)\}/3$$
$$+ 2n_5/4 + 2n_6/5 + 2n_7/6$$

$$\leqslant \sum_{j=2}^{7} \sum i_j n_d(i_2, i_3, \ldots, i_7)/(j-1)$$

$$= \sum n_d(i_2, i_3, \ldots, i_7) \sum_{j=2}^{7} i_j/(j-1). \tag{6.5}$$

We next show that if $n_d(i_2, i_3, \ldots, i_7) \neq 0$ then

$$\sum_{j=2}^{7} i_j/(j-1) \leqslant 1. \tag{6.6}$$

Let w be any vertex of type $(i_2, i_3, \ldots, i_7)$ having degree d. Let l be the minimum degree of the neighbours of w, then $i_j \neq 0$ implies $l \leqslant j$. Let v be a neighbour of w with $d(v) = l$, then $d^*(w) \geqslant d - l + 1$ since edge (v, w) is not eliminable. Therefore we have

$$\sum_{j=2}^{7} i_j \leqslant l - 1,$$

which implies (6.6).

By the definition we have

$$n_d = \sum n_d(i_2, i_3, \ldots, i_7). \tag{6.7}$$

Combining (6.5), (6.6) and (6.7), we have

$$n_d \geqslant 2n_2 + \{2r + 3(n_3 - r)\}/2$$
$$+ \{3s + 3t + 2u + 4(n_4 - s - t - u)\}/3 + 2n_5/4$$
$$+ 2n_6/5 + 2n_7/6,$$

which immediately yields the following.

$$(d-7)n_{d-1} + 2n_d \geqslant 4n_2 + 3n_3 + 2n_4 + n_5$$
$$+ 2(n_4 - s - t)/3 + (n_{d-1} - r - u)$$
$$+ \{2n_7/3 + (d-8)n_{d-1} - u/3\}. \tag{6.8}$$

The definition of s and t implies that

$$n_4 - s - t \geqslant 0. \tag{6.9}$$

If a vertex w of degree $d - 1$ is adjacent to a vertex v of degree 3, then all the neighbours of w except v have degree d. Therefore, among n_{d-1} vertices of degree $d - 1$, at most $n_{d-1} - r$ are adjacent to a vertex of degree 4, and furthermore each of such vertices is adjacent to at most two vertices of degree

4. Therefore, counting the number of edges with one end of degree 4 and the other end of degree $d - 1$, we have $t + 2u \leqslant 2(n_{d-1} - r)$, which implies

$$n_{d-1} - r - u \geqslant 0. \tag{6.10}$$

We now show that

$$2n_7/3 + (d - 8)n_{d-1} - u/3 \geqslant 0. \tag{6.11}$$

Suppose first that $d = 8$. Then, noting that $n_7 = n_{d-1}$ and using (6.10), one can easily verify (6.11). Suppose next that $d \geqslant 9$. Then from (6.10) we easily obtain $(d - 8)n_{d-1} - u/3 \geqslant 0$, implying (6.11).

Thus from (6.8)–(6.11) we have

$$(d - 7)n_{d-1} + (d - 6)n_d \geqslant 4n_2 + 3n_3 + 2n_4 + n_5,$$

which contradicts (6.1). Q.E.D.

The following is an immediate consequence of Lemma 6.2.

Theorem 6.3. [GNKLT85] *Algorithm ALCOLOR edge-colors a planar graph G with d colors if $d \geqslant 8$.*

6.3.2. Series-parallel graphs

A (simple) graph G is said to be *series-parallel* if G contains no K_4 as a subcontraction, that is, K_4 cannot be obtained from G by repeating the deletion or contraction of edges. The class of series-parallel graphs is a subclass of planar graphs, but large enough to include the class of outerplanar graphs. We have the following lemma.

Lemma 6.3. *Any series-parallel graph G whose maximum degree $d \geqslant 4$ has an eliminable edge.*

Proof. Since G is series-parallel, G has a vertex of degree at most two [Ore62]. Let S be the set of such vertices. Clearly $G' = G - S$ is also series-parallel. Therefore G' has a vertex w of degree at most two. Since the degree of w was at least three in G, a vertex $v \in S$ was adjacent with w in G. Thus we have $d^*(w) \leqslant 2$, so $d(v) + d^*(w) \leqslant 4$. Hence edge (v, w) is eliminable. Q.E.D.

The following is an immediate consequence of Lemma 6.3.

Theorem 6.4. *Algorithm ALCOLOR edge-colors a series-parallel graph G with d colors if d $\geq$ 4.*

A series-parallel graph G does not always contain an eliminable edge if $d = 3$. Therefore the direct application of ALCOLOR does not always produce a $q^*(G)$-coloring for the case $d = 3$. However we have the following lemma.

Lemma 6.4. *If a series-parallel graph G with d = 3 has no eliminable edge, then G has a triangle uvw such that d(v) = 2 and d(u) = d(w) = 3.*

Proof. Let v be any vertex of degree two, and let u and w be the neighbours of v. Since none of edges (v, u), (v, w), (u, v), and (w, v) is eliminable, we have $d(v) + d^*(u) \geq 4$, $d(v) + d^*(w) \geq 4$, and $d^*(v) = 2$. Hence $d(u) = d(w) = 3$ and $d^*(u) = d^*(w) = 2$. We shall show that u, v and w constitute a triangle, that is, u is adjacent to w. Suppose, contrary to the claim, that u is not adjacent to w for every vertex v of degree two. Then contract one of the two edges incident on v for every vertex v of degree two. The resulting graph G' has neither multiple edges nor vertices of degree two, so is a (simple) graph with maximum degree three. Since G is series-parallel, G' is also series-parallel and hence G' must contain a vertex of degree at most two, a contradiction.   Q.E.D.

Let G' be the graph obtained from G by contracting the triangle specified by Lemma 6.4 into a single vertex. G' is also series-parallel and has vertices two less than G. Clearly any 3-coloring of G' can be extended into a 3-coloring of G. Thus we have shown that $q^*(G) = 3$ for any series-parallel graph G with $d = 3$. Obviously $q^*(G) = d$ if $d \leq 2$ and G is not an odd cycle. These facts together with Theorem 6.4 imply the following theorem.

Theorem 6.5. *[GNKLT85] If a series-parallel graph G is not an odd cycle, then q*(G) = d.*

Using Lemma 6.4, we can easily modify algorithm ALCOLOR so that it would color any series-parallel graph G with $q^*(G)$ colors. Theorem 6.5 implies Fiorini's result that every outerplanar graph except odd cycles has an edge-coloring with d colors [Fio75].

6.3.3. Random graphs

Start with n distinguished (labelled) vertices, and choose every edge with a fixed probability p, $0 < p < 1$, independently of the choices of the other edges. The resulting graph is called a *random graph*. Almost every random graph has exactly one vertex w of maximum degree (see [Bol79, Theorem 9, pp. 135–136]). Let v be any neighbour of w, then $d(v) + d^*(w) \leq d$. Thus (v, w) is eliminable, and moreover $G - (v, w)$ has maximum degree one less than d. Hence algorithm ALCOLOR colors almost every random graph G with d colors.

6.4. Edge-coloring multigraphs

In this section we present an approximation algorithm MULTICOLOR, given by Nishizeki and Sato [NS84], which edge-colors any given multigraph G with at most $q(G) = \max\{p(G), \lfloor(5d(G) + 2)/4\rfloor\}$ colors. Throughout this section G denotes a given *multigraph* which may contain multiple edges but no self-loops. Denote by (v, w) an edge joining vertices v and w. Denote by $C(v, w)$ the set of colors assigned to the multiple edges joining vertices v and w. An edge colored c is called a *c-edge*. If all the edges of G except an edge $e = (v, w)$ are colored with q colors, $a \in M(v)$ and $b \in M(w)$, then the *ab*-alternating path between v and w, if any, is called an *ab-critical path*. Note that every critical path contains an odd number of vertices.

Algorithm MULTICOLOR is outlined as follows.

procedure MULTICOLOR(G);
begin $\{G = (V, E)\}$
   if $d(G) \leq 2$
   then color G with $q^*(G)$ colors by a trivial method
   else $\{d \geq 3\}$
     begin
       $q := \lfloor(5d + 2)/4\rfloor$; $\{q$ colors are currently available$\}$
       $G' := (V, \varnothing)$; $\{$empty subgraph of $G\}$
       for each $e \in E$ *do*
          begin
             $G' := G' + e$; $\{$add an uncolored edge $e = (v, w)\}$
             MULTIRECOLOR(G', e) $\{$update coloring$\}$
          end
     end
end.

Suppose that all the edges of subgraph G' except one edge $e = (v, w)$ have been colored with q colors. Then *procedure* MULTIRECOLOR(G', e) finds either a q-coloring or a $(q + 1)$-coloring of the whole of G' including e by recoloring some subgraphs of G' or introducing a new color; in the latter case q satisfies $q < p(G) \leqslant q(G)$. Whenever MULTIRECOLOR is executed, $d(G) \geqslant 3$ and hence the number q of currently available colors exceeds the maximum degree, that is, $q \geqslant \lfloor(5d + 2)/4\rfloor \geqslant d + 1$. Therefore every vertex of G' has at least one missing color, while both ends v and w of the uncolored edge e have at least two missing colors. Let color a be an arbitrary missing color of v, and let b be an arbitrary missing color of w.

In what follows we describe the details of MULTIRECOLOR, separating the following three cases:

Case 1: either v and w have a common missing color or G' has no ab-critical path.

In this case we execute the following procedure APATH.

procedure APATH(e, a, b):
begin
   if G' has no ab-critical path *then*
     begin
       $\{a \in M(v), b \in M(w)$, and Apath(v, a, b) does not end at $w\}$
       interchange the colors a and b of Apath(v, a, b)
     end;
   $\{v$ and w now have a common missing color$\}$
   assign a common missing color of v and w to the uncolored edge $e = (v, w)$;
   $\{$the resulting is a q-coloring of the whole of $G'\}$
end;

Case 2: the ab-critical path $Q = $ Apath(v, a, b) contains two vertices x and y having a common missing color c.

We reduce this case to Case 1 above. If x is followed by y on path Q, then one can destroy the critical path Q simply by recoloring edge (x, y) with color c. In the resulting q-coloring of G' there exists no more ab-critical path, so APATH is applicable. Even if x is not followed by y on path Q, one can destroy Q as follows: first make two successive vertices on Q have a common missing color by repeating the interchange of the colors of an alternating path, and then recolor the edge joining them with the common missing color. Thus in this case we execute the following procedure CPATH.

procedure CPATH(e, a, b):
begin

find two vertices x and y on Q having a common missing color c;
assume that v, x, y, w appear on Q in this order;
{possibly $v = x$ or $y = w$}
while x is not followed by y on Q *do*
    begin
        let x' be the next vertex of x on Q;
        let c' be any missing color of x';
        interchange colors of the cc'-alternating path $R = \text{Apath}(x', c', c)$;
        if R ends at x *then* $x := x'$ *else* $y := x'$;
        {new x and y have a common missing color c, and the distance between
        them decreases by at least one.}
    end
{Q has two successive vertices x and y of a common missing color c}
assign color c to the edge $e = (x, y)$ on Q;
{there is no ab-critical path}
APATH(e, a, b)
end;

We obtain the following lemma from the definition of $q(G)$.

Lemma 6.5. *Assume that* $d(G) \geqslant 3$, *and that all the edges of* G' *except* $e = (v, w)$ *are colored with* $q(G)$ *colors. Let* S *be any subset of* V *such that no two vertices of* S *have a common missing color.*
(a) *If* $v, w \in S$, *then* $|S| \leqslant 4$.
(b) *If* $v \in S$, *then* $|S| \leqslant 5$.

Proof. (a) Clearly $|M(v)|, |M(w)| \geqslant q - d + 1$, and $|M(x)| \geqslant q - d$ for every vertex $x \in V$ other than v and w. Since $\Sigma_{x \in S} |M(x)| \leqslant q$, we have

$$2(q - d + 1) + (|S| - 2)(q - d) \leqslant q,$$

which immediately yields

$$|S| \leqslant (q - 2)/(q - d).$$

Noting that $d \geqslant 3$ and $q \geqslant (5d + 2)/4 - \frac{3}{4}$, we easily have $|S| \leqslant 4$.
  (b) Similar to (a).   Q.E.D.

Lemma 6.5 implies that, if the ab-critical path Q contains five or more vertices, then Q contains two vertices having a common missing color among the first six vertices, and hence Case 2 necessarily occurs.

Case 3: the ab-critical path Q contains no two vertices of a common missing color.

By Lemma 6.5 Q contains exactly three vertices. Let $u \in V$ be the intermediate vertex of Q. An edge vu is colored b, and an edge uw is colored a. Denote by H the subgraph of G' induced by three vertices u, v and w, then $n(H) = 3$. An edge of G' is said to *leave H at vertex z* if it joins a vertex z in H and a vertex not in H. If $q < m(H) = m(H)/\lfloor n(H)/2 \rfloor$, then $q < p(G)$ and hence we can and must introduce a new color to color edge (v, w). Thus we may assume that $q \geqslant m(H)$. Since edge (v, w) in H is uncolored, there is a color c not used in the coloring of H. Since at most one of u, v, w misses color c, there are two or three edges leaving H and colored with the same color c. Then the coloring of G' can be modified so that an edge colored with a missing color of v or w leaves H. In the resulting coloring there is a critical path having five or more vertices, so CPATH is applicable. Thus in this case we execute the following procedure.

procedure THREE:
begin
  if $q < m(H)$ *then*
    begin
      $\{q < p(G)\}$
      assign a new color $q + 1$ to edge (v, w);
      $q := q + 1$
    end
  else {there are two or three edges leaving H and colored with the same color}
    begin
      if G' has no edge leaving H and colored with a missing color of u, v, or w
        then
        begin
          $\{G'$ has three c-edges leaving H, where c is not missing at any vertex of $H\}$
          let uu' be a c-edge leaving H; $\{u' \notin H\}$
          let f be any missing color of u; $\{f \neq a, b, c\}$
          let xx' be the final c-edge of Apath(u, f, c) leaving H' $\{x' \notin H$, and $x = u$, v, or $w\}$
          if $x = u$ *then* interchange the colors of Apath(u, f, c) $\{c \in M(u)$, and there is a c-edge leaving $H\}$
          else $\{x = v$ or $w\}$
            begin
              let g be any missing color of x such that $g \neq a, b$;

{since no *f*- or *g*-edge leaves *H*, Apath(*x*, *g*, *f*) does not leave *H* and ends at *u*};
interchange the colors of Apath(*x*, *g*, *f*);
{now *f*∈*M*(*x*)};
interchange the colors of Apath(*x*, *f*, *c*);
{*c* ∈ *M*(*x*), and there is a *c*-edge leaving *H*}
    end
  end;
{*G'* has two edges leaving *H* and colored with a missing color of a vertex of *H*}
if two edges colored with a missing color of *u* leaves *H*
then
  begin
    assign color *b* to the uncolored edge *e* = (*v*, *w*);
    erase color *b* from (*v*, *u*);
    interchange the names *u* and *w*
  end;
{two edges colored with a missing color of *v* or *w* leave *H*}
assume w.l.o.g. that color *c* is missing at *v*, and two *c*-edges leave *H*;
if there is no *bc*-critical path *then* APATH(*e*, *c*, *b*)
else {the *bc*-critical path contains at least five vertices, and hence CPATH is applicable}
CPATH(*e*, *c*, *b*)
  end
end;

Carefully reading the comments in the procedures, one can easily see that MULTIRECOLOR(*G'*, *e*) finds either a *q*- or (*q* + 1)-coloring of the whole *G'* including *e*; in the latter case necessarily *q* + 1 ≤ *q*(*G*). Therefore MULTICOLOR correctly edge-colors any multigraph *G* with at most *q*(*G*) colors.

We claim that the algorithm runs in O(*m*(*d* + *n*)) time. It suffices to show that one execution of MULTIRECOLOR can be done in O(*d* + *n*) time, since MULTICOLOR repeats MULTIRECOLOR *m* times. In what follows we show that the recoloring of each case can be done in O(*d* + *n*) time.

Case 1: Scanning the adjacency lists for vertices *v* and *w*, one can easily find sets *M*(*v*) and *M*(*w*) in O(*d*) time. Note that *q*(*G*) = O(*d*). Therefore we can know in O(*d*) time whether *v* and *w* have common missing color. Using the color lists for colors *a* and *b*, one can find Apath(*v*, *a*, *b*) in O(*n*) time, and hence know whether there exists an *ab*-critical path *Q*. Therefore one execution of APATH including the updating of the color lists can be done in $O(d + n)$ time.

Case 2: First consider the time required to decide whether Q contains two vertices x and y having a common missing color and to find them if they exist. If Q has exactly three vertices, then clearly it can be done in $O(d + n)$ time. Otherwise, by Lemma 6.5(b) Q necessarily contains two vertices of a common missing color among the first six vertices. Therefore it can be done in $O(d + n)$ time. Next consider the time required for the *while* statement in CPATH. Since one execution of the *while* statement decreases the distance between x and y at least one, the statement is executed at most four times. Clearly one execution of the statement can be done in $O(d + n)$ time. Therefore the statement spends $O(d + n)$ time in total. Thus CPATH can be done in $O(d + n)$ time.

Case 3: Clearly the time required by THREE, exclusive of the time by CPATH or APATH called there, is clearly $O(d + n)$. Thus the recoloring of Case 3 can be done in $O(d + n)$ time.

We have obtained the following theorem.

Theorem 6.6. *Algorithm MULTICOLOR edge-colors any multigraph G with $q(G)$ colors in $O(m(d + n))$ time and in $O(m)$ space, where $q(G) = \max\{p(G), \lfloor(5d(G) + 2)/4\rfloor\}$. In particular MULTICOLOR edge-colors G with $q^*(G)$ colors if $d(G) \le 2$.*

The following is an immediate consequence of Theorem 6.6.

Collorary 6.1. *Algorithm MULTICOLOR has a worst-case ratio of $\frac{4}{3}$.*

Proof. If either $d(G) \le 2$ or $q(G) = p(G)$, then clearly $q(G)/q^*(G) = 1$. If $d(G) \ge 3$ and $q(G) = \lfloor(5d + 2)/4\rfloor > p(G)$, then

$$q(G)/q^*(G) \le \lfloor(5d + 2)/4\rfloor/d \le \tfrac{4}{3}.$$

Clearly there exists a graph G such that $q(G)/q^*(G) = \frac{4}{3}$, the complete graph K_4, for example.   Q.E.D.

Furthermore the following is an immediate consequence of Holyer's result [Hol81].

Lemma 6.6. *If $P \neq NP$, then no polynomial-time approximation algorithm for edge-coloring multigraphs has a worst-case ratio less than $\frac{4}{3}$.*

Proof. By Theorem 6.1 every cubic (simple) graph can be edge-colored with three or four colors. Suppose that there exists a polynomial-time approximation algorithm with a worst-case ratio less than $\frac{4}{3}$. Then the problem of deciding whether a given cubic graph can be edge-colored with three colors is solvable in polynomial time: the answer to the decision problem is YES if and only if the algorithm produces a 3-coloring. Thus the problem belongs to the class P. However Holyer has shown that the problem is NP-complete [Hol81]. Hence $P = NP$, a contradiction.   Q.E.D.

Since the event $P \neq NP$ is very plausible, one cannot hope for an algorithm with a better worst-case ratio.

CHAPTER 7

INDEPENDENT VERTEX SETS

7.1. Introduction

A set of vertices in a graph is *independent* if no two vertices in the set are adjacent. The *maximum independent set problem*, in which one would like to find a maximum independent set in a given graph, is *NP*-complete, and still remains so even for the class of planar graphs. Thus there is no realistic hope of being able to design a polynomial-time algorithm for exactly solving the problem. Therefore it is expected to provide an efficient approximation algorithm, which does not necessarily find the maximum independent set but finds a large one.

An approximation algorithm is often evaluated by the *worst-case ratio*: the smallest ratio of the size of an approximation-solution to the size of a maximum solution, where the ratio is taken over all problem instances. It is known that if a polynomial time algorithm existed with a constant worst-case ratio > 0 for the maximum independent set problem on general graphs, then one could design a polynomial time algorithm with any constant worst-case ratio < 1 [HU79]. This fact does not imply that there exist no polynomial time approximation algorithms with a constant worst-case ratio > 0 for the problem on a special class of graphs, such as planar graphs. In fact, Lipton and Tarjan [LT80] have given an $O(n \log n)$ time approximation algorithm with a worst-case ratio $1 - O(1/(\log \log n)^{1/2})$, asymptotically tending to 1 as $n \to \infty$ (see Chapter 9). Such a ratio is called an "asymptotic worst-case ratio". On the other hand, some approximation algorithms have an "absolute worst-case ratio", which does not depend on the size n of a graph. For example, the 4-coloring algorithm, derived from the proof of the four-color theorem, immediately yields an approximation algorithm for the problem with the worst-case ratio $\frac{1}{4}$: Simply output the largest class of vertices colored with the same color. On the other hand the 5-coloring algorithm in Chapter 5 achieves the absolute worst-case ratio $\frac{1}{5}$. Moreover the algorithm of Albertoson or Chiba et al. guarantees the worst-case ratio $\frac{2}{3}$ for the problem [Alb74, CNS83].

In Section 7.2 we present an $O(n \log n)$ time approximation algorithm with absolute worst-case ratio $\frac{1}{2}$, due to Chiba, Nishizeki and Saito [CNS82]. For a given planar graph of any number n of vertices, the algorithm finds, in $O(n \log n)$ time, an independent vertex set that is necessarily larger than half a maximum independent set. The algorithm is based on the following ideas:

(i) we can reduce the problem on a general planar graph to one on a planar graph having no vertices of degree 4 or less; such a graph cannot have a large independent set (see Theorem 1.6); and

(ii) we can design an $O(n \log n)$ time "on-line" algorithm to execute any sequence of vertex-identifications and edge-deletions of a given planar graph. (An *on-line* algorithm is allowed only to execute an instruction without looking at any following instruction.)

Finally in Section 7.3 we briefly discuss another approximation algorithm recently given by Baker [Bak83].

7.2. Approximation algorithm

We denote by $I(G)$ an independent set of G and $I^*(G)$ a maximum independent set of G. The graphs with labelled vertices v, v_1, v_2, v_3 and v_4 depicted in Fig. 7.1(a), (b) and (c) are denoted by H_4, H_4' and H_4'', respectively, which will be referred to in the algorithm. We have the following lemma on them.

Lemma 7.1. *Let a planar graph* $G = (V, E)$ *contain a vertex* v *of degree 4 with* $N(v) = \{v_1, v_2\, v_3, v_4\}$, *and let* H *be a subgraph of* G *induced by* $\{v\} \cup N(v)$. *Then, renaming vertices in* $N(v)$ *if necessary, we can assume that either*
(i) H *is isomorphic with* H_4; *or*
(ii) H *contains* H_4' *or* H_4'' *as a subgraph, and moreover* $(v_1, v_3) \notin E$.

Proof. Suppose that H is not isomorphic with H_4, then there exists at least one edge of both ends in $N(v)$, say (v_2, v_4). If $(v_1, v_3) \notin E$, then (ii) holds with respect to H_4'. If $(v_1, v_3) \in E$, then, renaming vertices in $N(v)$, we may assume that H contains H_4'' as a subgraph. Since G is planar, $(v_1, v_3) \notin E$ or $(v_2, v_4) \notin E$. Thus renaming vertices again if necessary, we have (ii) with respect to H_4''. Q.E.D.

The algorithm uses a simple recursion. The outline is as follows: from an original planar graph G we construct a planar graph G' smaller than G by deleting a vertex v of minimum degree together with some of its neighbours

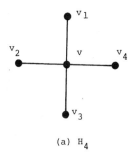

(a) H_4

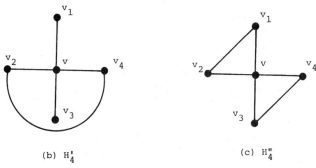

(b) H_4' (c) H_4''

Fig. 7.1. Graphs H_4, H_4' and H_4''.

and adding some edges; find an independent set $I(G')$ of G' larger than half a maximum independent set $I^*(G')$ of G' by recursively applying the algorithm; and form an independent set $I(G)$ of G by adding one or two of the deleted vertices to $I(G')$ so that $I(G)$ is larger in size than half a maximum independent set $I^*(G)$ of G. The method to construct G' from G varies with $d(v)$ and the structure of a subgraph of G induced by $\{v\} \cup N(v)$. We illustrate some pairs of G and G' in Figs. 7.2–7.5. In the algorithm, $G = (V, E)$ denotes a graph currently processed and $G' = (V', E')$ a graph reduced from G. The algorithm is as follows.

procedure INDEPENDENT-SET;
  procedure REDUCE($G, I(G)$);
    begin
      if $V = \varnothing$
        then $I(G) := \varnothing$
        else $\{V \neq \varnothing\}$

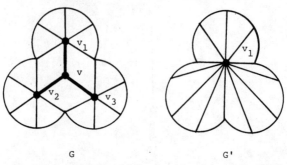

Fig. 7.2. G and G' where $d_m(G) = 3$ and $(v_1, v_2), (v_2, v_3), (v_3, v_1) \notin E$.

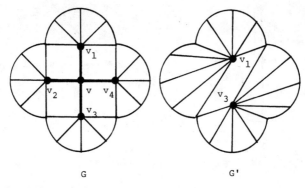

Fig. 7.3. G and G' where $d_m(G) = 4$ and the subgraph H of G induced by $\{v\} \cup N(v)$ is isomorphic with H_4.

```
begin
    let v be a vertex of minimum degree;
    case d(v) of
    ≤ 2: {d(v) ≤ 2}
      begin
          G' := G − {v} ∪ N(v);
          REDUCE(G', I(G'));
          I(G) := I(G') + {v}
      end;
        3 : {d(v) = 3} DEGREE3;
        4 : {d(v) = 4} DEGREE4;
        5 : {d(v) = 5} DEGREE5
    end
end
```

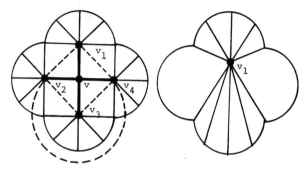

G   G'
Fig. 7.4. *G* and *G'* where $d_m(G) = 4$ and *H* contains H'_4 or H''_4 as a subgraph.

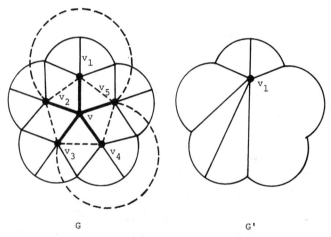

G   G'
Fig. 7.5. *G* and *G'* where $d_m(G) = 5$.

  end
begin
  embed a given planar graph *G* in the plane;
  REDUCE(*G*, *I*(*G*))
end.

procedure DEGREE3;
  begin
    let $N(v) = \{v_1, v_2, v_3\}$;

if $(v_1, v_2), (v_2, v_3), (v_3, v_1) \notin E$
   then {See Fig. 7.2}
     begin
       let G' be the graph obtained from $G - \{v, v_2, v_3\}$ by joining v_1 to all
       the vertices which were adjacent to v_2 or v_3 in G;
       {G' is the graph obtained from G by contracting the three edges
       incident to v}
       REDUCE$(G', I(G'))$;
       if $v_1 \in I(G')$
         then $I(G) := I(G') + \{v_2, v_3\}$
         else $I(G) := I(G') + \{v\}$
     end
       else {$(v_1, v_2), (v_2, v_3)$ or $(v_3, v_1) \in E$}
         begin
           $G' := G - \{v\} \cup N(v)$;
           REDUCE$(G', I(G'))$;
           $I(G) := I(G') + \{v\}$
         end
  end;

procedure DEGREE4;
  begin
    let H be a subgraph of G induced by $\{v\} \cup N(v)$;
    if H is isomorphic with H_4
      then {See Fig. 7.3}
        begin
          let vertices in $N(v) = \{v_1, v_2, v_3, v_4\}$ be arranged cyclically counter-
          clockwise about v in the plane embedding of G;
          let G' be the graph obtained from $G - \{v, v_2, v_4\}$ by joining v_1 to all
          vertices which were adjacent to v_2 and joining v_3 to all vertices
          which were adjacent to v_4 in G;
          REDUCE$(G', I(G'))$;
          if $v_1, v_3 \notin I(G')$ *then* $I(G) := I(G') + \{v\}$;
          if $v_1 \in I(G')$ *then* $I(G) := I(G') + \{v_2\}$;
          if $v_1 \notin I(G')$ and $v_3 \in I(G')$ *then* $I(G) := I(G') + \{v_4\}$
        end
        else {H contains either H'_4 or H''_4 as a subgraph (See Fig. 7.4).}
         begin
           label vertices in $N(v) = \{v_1, v_2, v_3, v_4\}$ so that H'_4 or H''_4 is a subgraph
           of the labeled graph H, and $(v_1, v_3) \notin H$;

let G' be the graph obtained from $G - \{v, v_2, v_3, v_4\}$ by joining v_1 to
all vertices which were adjacent to v_3 in G;
REDUCE(G', $I(G')$);
 if $v_1 \in I(G')$ *then* $I(G) := I(G') + \{v_3\}$
      else $I(G) := I(G') + \{v\}$
  end
end;

procedure DEGREE5;
  begin
   {See Fig. 7.5}
   let vertices in $N(v) = \{v_1, v_2, v_3, v_4, v_5\}$ be arranged cyclically counterclockwise about v in the plane embedding of G;
   {since G is planar, $\{v_1, v_2, \ldots, v_5\}$ does not induce K_5}
   wlog assume $(v_1, v_3) \notin E$;
   let G' be the graph obtained from $G - \{v, v_2, v_3, v_4, v_5\}$ by joining v_1 to all
   vertices which were adjacent to v_3;
   REDUCE(G', $I(G')$);
   if $v_1 \in I(G')$ *then* $I(G) := I(G') + \{v_3\}$
      else $I(G) := I(G') + \{v\}$
  end;

The following theorem holds on the correctness of the algorithm.

Theorem 7.1. *For any planar graph $G = (V, E)$, the algorithm finds an independent set of G with worst-case ratio $> \frac{1}{2}$.*

We first present the following result before establishing Theorem 7.1.

Theorem 7.2. *For any planar graph $G = (V, E)$ with n vertices, the algorithm
finds an independent set of at least $n/5$ vertices.*

Proof. We proceed by induction on the number n of vertices of G. One can
easily see that for any planar graph having $n \leq 4$ vertices, our algorithm always
finds an independent vertex set of at least one vertex. Hence our claim is true
for such a graph.

As the inductive hypothesis we assume that the claim holds for all planar
graphs having less than n vertices.

Let $G = (V, E)$ be a planar graph having $|V| = n \, (\geq 5)$ vertices. We denote
the minimum degree of G by $d_m(G)$. Let v be a vertex of minimum degree in G.
Since G is planar, the degree of v is 5 or less by Corollary 1.4. The algorithm

constructs from G a new planar graph $G' = (V', E')$ of fewer vertices than G, and recursively finds an independent set $I(G')$ of G'. We have $|I(G')| \geqslant |V'|/5$ by the inductive hypothesis. Although the construction of G' varies with the degree of v, our algorithm always adds at least one vertex to $I(G')$ to form an independent set $I(G)$ of G, so that $|I(G)| \geqslant |I(G')| + 1$. Clearly $|V| \leqslant |V'| + 5$. Combining these three inequalities, we have $|I(G)| \geqslant |V|/5$.   Q.E.D.

We are now ready to prove Theorem 7.1.

Proof of Theorem 7.1. Since the set $I(G)$ found by the algorithm is clearly independent in G, it is sufficient to prove that $|I(G)|/|I^*(G)| > \frac{1}{2}$. We proceed by induction on the number n of vertices of G. For any planar graph G with $n \leqslant 4$ vertices, we can easily verify that $|I(G)|/|I^*(G)| > \frac{1}{2}$.

As the inductive hypothesis, we assume that our claim is true for all planar graphs having less than n vertices.

Now suppose that G is a planar graph with n vertices, $n \geqslant 5$. Since G is planar, $d_m(G) \leqslant 5$. If $d_m(G) = 5$, we have $|I(G)| \geqslant n/5$ by Theorem 7.2, and $|I^*(G)| < 2n/5$ by Theorem 1.6. Therefore $|I(G)|/|I^*(G)| > \frac{1}{2}$, as desired. Thus we can assume $d_m(G) \leqslant 4$. Let G' be the graph constructed from G by our algorithm; the construction method varies with $d_m(G)$. Since G' has fewer vertices than G, we have $|I(G')|/|I^*(G')| > \frac{1}{2}$ by the inductive hypothesis. Our algorithm always adds at least one vertex to $I(G')$ to form $I(G)$, so that $|I(G)| \geqslant |I(G')| + 1$. Hence, we shall show that $|I^*(G)| \leqslant |I^*(G')| + 2$ which, together with the two inequalities above, leads to the desired result $|I(G)|/|I^*(G)| > \frac{1}{2}$.

Let v, v_1, v_2, v_3 and v_4 be vertices of G defined in the algorithm. We consider three cases depending on $d_m(G)$.

Case 1: either $d_m(G) \leqslant 2$ or $d_m(G) = 3$ and G has at least one of edges (v_1, v_2), (v_2, v_3) and (v_3, v_1).

In this case $G' = G - \{v\} \cup N(v)$. Since $I^*(G) - \{v\} \cup N(v)$ is independent in G', and $|(\{v\} \cup N(v)) \cap I^*(G)| \leqslant 2$, we have

$$|I^*(G')| \geqslant |I^*(G) - \{v\} \cup N(v)| \geqslant |I^*(G)| - 2,$$

as desired.

Case 2: $d_m(G) = 3$ and $(v_1, v_2), (v_2, v_3), (v_3, v_1) \notin E$. (See Fig. 7.2.)

In this case G' is the graph obtained from $G - S$, $S = \{v, v_2, v_3\}$, by joining v_1 to all the vertices which were adjacent to v_2 or v_3 in G. Clearly $|I^*(G) \cap S| \leqslant 2$. We consider two subcases depending on $|I^*(G) \cap S|$ as follows.

Subcase 2-1: $|I^*(G) \cap S| = 0$ *or* 1. Clearly $I^*(G) - \{v_1\} \cup S$ is independent in G'. Therefore $|I^*(G')| \geqslant |I^*(G)| - 2$, as desired.

Subcase 2-2: $|I^*(G) \cap S| = 2$. In this case $I^*(G) \cap S = \{v_2, v_3\}$, and hence $(N(v_2) \cup N(v_3)) \cap I^*(G) = \varnothing$. Therefore $I^*(G) - S$ is independent in G'. Thus $|I^*(G')| \geqslant |I^*(G)| - 2$.

Case 3: $d_m(G) = 4$.

Let H be the subgraph of G induced by $\{v\} \cup N(v)$. Then by Lemma 7.1, we can assume, without loss of generality, that either (a) H is isomorphic with graph H_4, or (b) H contains a subgraph isomorphic with graph H_4' or H_4'' (H_4, H_4' and H_4'' are depicted in Fig. 7.1).

First assume that H is isomorphic with H_4. Let G' be the graph obtained from $G - S$ as in a manner described in the algorithm, where $S = \{v, v_2, v_4\}$. (See G and G' in Fig. 7.3.) Since clearly $|I^*(G) \cap S| \leqslant 2$, we consider three subcases.

Subcase 3-1: $|I^*(G) \cap S| = 0$. Since $I^*(G) - \{v_1, v_3\}$ is independent in G', $|I^*(G')| \geqslant |I^*(G)| - 2$.

Subcase 3-2: $|I^*(G) \cap S| = 1$. Suppose first that $I^*(G) \cap S = \{v\}$. Then $v_1, v_3 \notin I^*(G)$, and hence $I^*(G) - \{v\}$ is independent in G'. Thus $|I^*(G')| \geqslant |I^*(G)| - 1$. Suppose next that $I^*(G) \cap S = \{v_2\}$ or $\{v_4\}$. We can assume without loss of generality that $I^*(G) \cap S = \{v_2\}$. Since $N(v_2) \cap I^*(G) = \varnothing$, $I^*(G) - \{v_2, v_3\}$ is independent in G', and hence $|I^*(G')| \geqslant |I^*(G)| - 2$.

Subcase 3-3: $|I^*(G) \cap S| = 2$. In this case $I^*(G) \cap S$ must be $\{v_2, v_4\}$. Since $N(v_2) \cap I^*(G) = \varnothing$ and $N(v_4) \cap I^*(G) = \varnothing$, $I^*(G) - \{v_2, v_4\}$ is independent in G', so $|I^*(G')| \geqslant |I^*(G)| - 2$.

Next assume that H contains H_4' or H_4'' as a subgraph. Let G' be the graph obtained from $G - S$ as in the manner described in the algorithm, where $S = \{v, v_2, v_3, v_4\}$. (See G and G' in Fig. 7.4.) Since $|I^*(G) \cap S| \leqslant 2$, we consider two subcases.

Subcase 3-4: $|I^*(G) \cap S| = 0$ *or* 1. Since $I^*(G) - (\{v_1\} \cup S)$ is independent in G', $|I^*(G')| \geqslant |I^*(G)| - 2$.

Subcase 3-5: $|I^*(G) \cap S| = 2$. Suppose first that H contains H_4' as a subgraph, then $I^*(G) \cap S$ must be either $\{v_2, v_3\}$ or $\{v_3, v_4\}$. Since $N(v_3) \cap I^*(G) = \varnothing$ in either case, $I^*(G) - S$ is independent in G', so $|I^*(G')| \geqslant |I^*(G)| - 2$. Suppose next that H contains H_4'' as a subgraph, then $I^*(G) \cap S$

must be either $\{v_2, v_3\}$ or $\{v_2, v_4\}$. Since $v_1 \notin I^*(G)$ in either case, $I^*(G) - S$ is independent in G', so $|I^*(G')| \geqslant |I^*(G)| - 2$.

Thus the proof is completed.   Q.E.D.

The bound on the worst-case ratio given in Theorem 7.1 is sharp in the sense that there exist infinitely many graphs for which our algorithm possibly realizes the bound. We illustrate an example of these graphs G in Fig. 7.6. For the graph G the algorithm finds, in the worst case, the set of all black vertices as an independent set $I(G)$. Clearly $|I(G)| = (|I^*(G)| + 1)/2$.

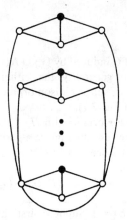

Fig. 7.6. An illustrating graph.

Finally we establish the following theorem on the running time of the algorithm.

Theorem 7.3. *The algorithm spends at most* O($n \log n$) *time on a planar graph* $G = (V, E)$.

We first give data structures appropriate for graph alterations and then present some results on the time required for a sequence of alterations.

Let u and v be two vertices of a graph $G = (V, E)$. Remember that identification of vertices u and v is denoted by $\langle u, v \rangle$. The algorithm frequently uses this operation to construct G' from G. As a data structure to represent a graph G, we use an adjacency matrix A together with adjacency lists Adj. The $u - v$ element $A(u, v)$ of A is 1 if and only if vertex v is adjacent to vertex u. $A(u, v)$ also has a pointer to the element "v" in Adj(u) if $(u, v) \in E$. It is an easy exercise to avoid the O(n^2) time necessary to initialize the matrix A. (See

[AHU74, Ex. 2.12].) In addition to A and Adj, we use two arrays D and DP together with six lists DLIST(i), $0 \le i \le 5$, so that one can find a vertex of minimum degree in a constant time. An element $D(v)$ of array D contains the value of the degree $d(v)$, $v \in V$. DLIST(i) contains all the vertices of degree i, $0 \le i \le 5$. $DP(v)$ has a pointer to an element "v" in DLIST($d(v)$) if $d(v) \le 5$. Fig. 7.7 illustrates such a data structure. Using these data structures we have the following results on graph alterations.

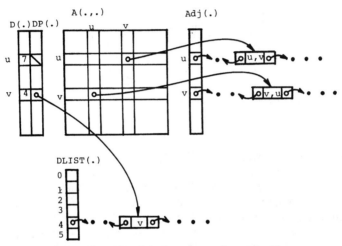

Fig. 7.7. Illustration of the data structure for our algorithm.

Lemma 7.2. *Let $G = (V, E)$ be a graph, and let s be any sequence of edge-deletions on G. There exists an on-line algorithm to execute s in $O(m)$ time.*

Proof. In order to delete an edge, say (u, v), of G, we substitute 0 to elements $A(u, v)$ and $A(v, u)$, and delete elements u from list Adj(v) and v from Adj(u). One can directly access these elements in the lists via the pointers in the corresponding elements in A, and one can delete an element in the list in a constant time since it is doubly linked. Thus the execution of each edge-deletion requires a constant time. Of course s contains at most m edge-deletions. Hence the total amount of time required to execute s is at most $O(m)$.   Q.E.D.

Lemma 7.3. *Let $G = (V, E)$ be a graph, and let s be any sequence of identifica-*

tions on G (see Fig. 7.8). There exists an on-line algorithm to execute s in at most $O(\min\{m \log n, n^2\})$ *time.*

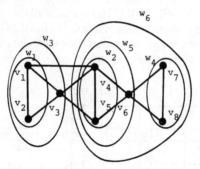

Fig. 7.8. A sequence of identifications $s = \langle v_1, v_2 \rangle \ \langle v_4, v_5 \rangle \ \langle w_1, v_3 \rangle \langle v_7, v_8 \rangle \ \langle w_2, v_6 \rangle \ \langle w_4, w_5 \rangle$.

Proof. Let $s = \langle u_1, v_1 \rangle \langle u_2, v_2 \rangle \cdots \langle u_l, v_l \rangle$, where $l \leq n - 1$. Let G_i, $i = 1, 2, \ldots, l + 1$, be the graph obtained from G by the first $i - 1$ identifications.

In order to update the data structure for G_{i+1} from that of G_i, one shall modify some adjacency lists and entries of A. Assume $d(G_i, u_i) \leq d(G_i, v_i)$. Scanning all the elements x of list $\text{Adj}(u_i)$ (i.e., all the vertices x adjacent to u_i), we delete element u_i from $\text{Adj}(x)$, and add element x to $\text{Adj}(v_i)$ if $x \neq v_i$ and x is not in $\text{Adj}(v_i)$. Note that we consider only a simple graph, which has neither multiple edges nor loops. In our data structure element x in $\text{Adj}(v_i)$ can be directly accessed via the pointer in $A(v_i, x)$. We also modify some entries of A appropriately according to the above modification of lists. Then we remove the list $\text{Adj}(u_i)$, and regard the updated list $\text{Adj}(v_i)$ as a list $\text{Adj}(w_i)$ for a vertex w_i added to G_i in place of u_i and v_i. Hence the time required to execute the identification $\langle u_i, v_i \rangle$ is proportional to $d(G_i, u_i)$. Therefore the total amount of time $T(s)$ required for s satisfies

$$T(s) = O \left(\sum_{1 \leq i \leq l} \min\{d(G_i, u_i), d(G_i, v_i)\} \right).$$

Since $l \leq n - 1$ and $d(G_i, u_i), d(G_i, v_i) \leq n$ for $1 \leq i \leq l$, clearly $T(s)$ is at most $O(n^2)$. Therefore we shall show that $T(s)$ is at most $O(m \log n)$.

Graphs G_i, $1 \leq i \leq l + 1$, are all *simple*, that is, the newly produced multiple edges and loops are deleted in them. Consider multigraphs G_i^*, $1 \leq i \leq l + 1$, corresponding to G_i, in which the multiple edges and loops produced by identifications are not deleted. Define the degree of a vertex in a multigraph to be the number of edges incident on the vertex, counting loops twice. Then obviously $d(G_i, u_i) \leqslant d(G_i^*, u_i)$ and $d(G_i, v_i) \leqslant d(G_i^*, v_i)$, and hence,

$$\min\{d(G_i, u_i), d(G_i, v_i)\} \leqslant \min\{d(G_i^*, u_i), d(G_i^*, v_i)\}).$$

Although the identification $\langle u_i, v_i \rangle$ spends only the cost equal to the left-hand side, we apportion the cost equal to the right-hand side uniformly among the (directed) edges incident to a vertex, either u_i or v_i, say u_i, having smaller degree in G_i^*. The surrogate w_i of u_i in G_{i+1}^* has degree at least twice as large as u_i in G_i^*. The degree of a vertex is nondecreasing, and the degree of a vertex in G_{i+1}^* is clearly at most $2m$. Therefore no (directed) edge in the multigraph receives more than log $2m$ costs. Therefore the total cost is at most $2m$ log-$2m = O(m \log n)$. This completes the proof.   Q.E.D.

Theorem 7.4. *For any sequence s of identifications and edge-deletions of a graph $G = (V, E)$, there exists an on-line algorithm to execute s in at most $O(\min\{m \log n, n^2\})$ time.*

Proof. By Lemma 7.2 the time required to execute all the edge-deletions in s is $O(m)$. Hence we shall show that the time T required to execute all the remaining identifications is at most $O(\min\{m \log n, n^2\})$. Let s' be a subsequence of s consisting of all identifications, and consider the time T' required for s'. Let G_i be the graph resulted from G after the first i terms of s, including j identifications. Let G_j' be the graph as a result of G after the first j identifications of s'. Since G_i is a subgraph of G_j', $d(G_i, v) \leqslant d(G_j', v)$ for every vertex v of G_i. Noting this fact, we have $T \leqslant T'$. Therefore by Lemma 7.3 $T \leqslant O(\min\{m \log n, n^2\})$, which establishes our claim.   Q.E.D.

We are now ready to prove Theorem 7.3.

Proof of Theorem 7.3. One can embed a planar graph G in the plane in $O(n)$ time, using the algorithm in Chapter 3.

In the algorithm INDEPENDENT-SET, the problem to find an independent set of G is reduced to that of a smaller graph G', which can be obtained from G by deleting a vertex of minimum degree together with some other modifications. Since we always keep all the vertices of degree i ($i = 0, 1, \ldots, 5$) in list DLIST(i) (see Fig. 7.7), we can find a vertex v of minimum degree (necessarily at most 5) in a constant time. Note that we can delete an element from the doubly linked list DLIST, or insert an element into it, in a constant time, since we can directly access an element v of DLIST via the pointer $DP(v)$. Although the method to form $I(G)$ from $I(G')$ varies with the type of a subgraph of G

induced by $\{v\} \cup N(v)$, one can recognize the type in a constant time. Therefore, given $I(G')$ obtained by recursively applying the algorithm, we can form $I(G)$ in a constant time. Since the graph G' reduced from G has fewer vertices than G, such a reduction occurs at most n times. Therefore the total amount of time required to execute these operations is at most $O(n)$. All the remaining operations, which are used to construct G' from G, are regarded as a sequence of three kinds of operations: identification, edge-deletion, and deletion of an isolated vertex. For example, G' of Fig. 7.3 can be constructed from G by a sequence of four edge-deletions, two identifications, and one deletion of an isolated vertex. Theorem 7.4 implies that any sequence of the first two kinds of operations can be executed within the desired time, while one can easily see that any sequence of deletions of an isolated vertex is executed in $O(n)$ time. Thus the proof is completed. Q.E.D.

7.3. Baker's algorithm

Recently Baker gave an elegant approximation algorithm for the maximum independent set problem on planar graphs, which attains the worst-case ratio $k/(k + 1)$ and runs in $O(8^k kn)$ time for any positive integer k. Thus her algorithm realizes the worst-case ratio of both types; absolute and asymptotic. For example, letting $k = 1$ we can get a linear time algorithm having the absolute worst-case ratio $\frac{1}{2}$, while letting $k = \log \log n$ we can get an $O(n(\log n)^3 \log \log n)$ time algorithm with the asymptotic worst-case ratio of $\log \log n/(1 + \log \log n)$ tending to 1. The algorithm is outlined as follows:

(0) choose k as an appropriate integer.

(1) consider a plane embedding of a given graph G and define the level of a vertex as follows: the vertices on the outer boundary have *level* 1; the vertices on the outer boundary of the subgraph obtained by deleting the level $i - 1$ vertices have *level* i;

(2) for each i, $0 \leq i \leq k$, do the following: construct graph $G_i = G - V_i$, where V_i is the set of all the vertices of level i (mod $k + 1$) (all the components of G_i are "k-outerplanar", that is, it has an embedding having no vertex of level $> k$); find a maximum independent set $I^*(G_i)$ of G_i by taking union of those of the components;

(3) output the largest $I^*(G_i)$ as the solution of G.

A maximum independent set $I^*(G)$ of G is partitioned into $k + 1$ subsets $\{I^*(G) \cap V_i | i = 0, 1, \ldots, k\}$. Clearly the smallest one contains at most $1/(k + 1)$ of $I^*(G)$, and $|I^*(G) - V_i| \leq |I^*(G_i)|$. Therefore the largest $I^*(G_i)$ satisfies $|I^*(G_i)| \geq k|I^*(G)|/(k + 1)$. Thus the algorithm has the worst-case ratio $k/(k + 1)$. Furthermore, applying the dynamic programming to each component of G_i which is k-outerplanar, one can find $I^*(G_i)$ in $O(8^k n)$ time. So the entire algorithm spends at most $O(8^k kn)$ time.

It seems that Baker's algorithm always find an independent set $I(G)$ such that $|I(G)|$ is close to $k|I^*(G)|/(k + 1)$. Although the algorithm of Chiba et al. has the worst-case ratio of $\frac{1}{2}$, many computational experiments show that it usually finds an independent set close to the optimal one. Thus the algorithm seems to be sufficient for many practical applications.

CHAPTER 8

LISTING SUBGRAPHS

8.1. Introduction

The problems involved in listing certain kinds of subgraphs of a graph arise in many practical applications [AM70, BE82, IR78, NAW83, RT75, TIAS77]. In this chapter we present a simple strategy for edge-searching a graph due to Chiba and Nishizeki [CN85a], which is useful for the various subgraph listing problems, especially for planar graphs. We choose a vertex v in a graph and scan the edges of the subgraph induced by the v's neighbours to find the pattern subgraphs containing v. The feature of the strategy is to repeat the searching above for each vertex v in decreasing order of degree and to delete v after v is processed so that no duplication occurs. It will be shown in the succeeding section that the procedure above requires $O(a(G)m)$ time. Throughout this chapter $a(G)$ is the *arboricity* of G, that is, the minimum number of edge-disjoint spanning forests into which G can be decomposed. For example the graph G in Fig. 8.1(a) can be decomposed into two forests as in (b), and hence $a(G) = 2$. We use the rather unfamiliar graph invariant $a(G)$ as a parameter in bounding the running time of algorithms.

The strategy yields simple algorithms for the problems to list certain kinds of subgraphs of a graph [CN85a]: "triangle", "quadrangle", "clique of a fixed order", and "maximal clique."

In Section 8.2 we give an upper bound on $a(G)$ for a general graph $G : a(G) \leqslant \lceil (2m + n)^{1/2}/2 \rceil$, which implies $a(G) \leqslant O(m^{1/2})$ for a connected graph G. In Section 8.3 we give a simple algorithm which lists all the triangles in an arbitrary graph G in $O(a(G)m)$ time. In Section 8.4 we present an $O(a(G)m)$ time algorithm for finding all the quadrangles (i.e. C_4) in G, which does not actually list C_4 but finds a representation of all the C_4. If G is planar, $a(G) \leqslant 3$, so these two algorithms run in linear time for planar graphs. Because of the bound on $a(G)$, they run in at most $O(m^{3/2})$ time for general graphs.

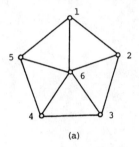

(a)

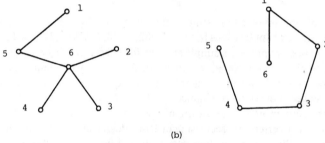

(b)

Fig. 8.1. (a) A graph and (b) trees.

Finally in Section 8.5 we present a core part of an algorithm for listing all the maximal cliques in G in $O(a(G)m)$ time per maximal clique. All the algorithms require at most linear space.

8.2. Arboricity and efficient edge-searching

We first define some terms. A *triangle* in a graph is a cycle of length three (i.e. C_3), in other words, a *clique* (or complete subgraph) of three vertices (i.e. K_3). A *maximal clique* is a maximal complete subgraph in a graph. Remember that $\lceil x \rceil$ is the smallest integer not less than x.

We next present two results; the first is concerned with the arboricity of a graph and the other with the time required by scanning edges with our strategy.

Lemma 8.1. [CN85a] *Let G be a graph, then*

(a) $a(G) \leqslant \lceil (2m+n)^{1/2}/2 \rceil$;   (8.1)
(b) $a(G) \leqslant \lceil n/2 \rceil$; *and*
(c) $a(G) \leqslant 3$ *if G is planar.*

Proof. (a) Nash-Williams [Nas61] showed that

$$a(G) = \max_{H \subseteq G} \lceil q/(p-1) \rceil, \qquad (8.2)$$

where H runs over all nontrivial subgraphs of G, p is the number of vertices and q the number of edges of H. Suppose that the maximum in the right-hand side of (8.2) is achieved by a subgraph H having p vertices and q edges. Let k be the number of edges in a clique with p vertices, that is, $k = p(p-1)/2$. There are two cases to consider.

Case 1: $k \leqslant m$.

$$
\begin{aligned}
a(G) &= \lceil q/(p-1) \rceil \\
&\leqslant \lceil k/(p-1) \rceil \\
&= \lceil p/2 \rceil \\
&= \lceil (2k+p)^{1/2}/2 \rceil \\
&\leqslant \lceil (2m+n)^{1/2}/2 \rceil.
\end{aligned}
$$

Case 2: $k \geqslant m$.

$$
\begin{aligned}
a(G) &= \lceil q/(p-1) \rceil \\
&\leqslant \lceil m/(p-1) \rceil \\
&\leqslant \lceil \{mk/(p-1)^2\}^{1/2} \rceil \\
&= \lceil \{m(p-1)+m)/2(p-1)\}^{1/2} \rceil \\
&\leqslant \lceil \{m/2 + k/2(p-1)\}^{1/2} \rceil \\
&= \lceil (2m+p)^{1/2}/2 \rceil \\
&\leqslant \lceil (2m+n)^{1/2}/2 \rceil.
\end{aligned}
$$

(b) Immediate from (8.2).
(c) If G is planar, (8.2) together with Corollary 1.1 imply that $a(G) \leqslant \max_{H \subseteq G} \lceil (3p-3)/(p-1) \rceil = 3$.   Q.E.D.

Since $a(K_n) = \lceil n/2 \rceil = \lceil (2m+n)^{1/2}/2 \rceil$ where $m = n(n-1)/2$, there exist an infinite number of graphs attaining the upper bound in (8.1). In this sense the bound is the best possible. It should be noted that $a(G) = O(1)$ for a large class of graphs including (i) planar graphs, (ii) graphs of bounded genus, and (iii) graphs of bounded maximum degree.

Lemma 8.2. [CN85a] *Let* $G = (V, E)$ *be a graph, then*

$$\sum_{(u,v) \in E} \min\{d(u), d(v)\} \leq 2a(G)m.$$

Proof. Let F_i $(1 \leq i \leq a(G))$ be the edge-disjoint spanning forests of G such that $E = \bigcup_{1 \leq i \leq a(G)} E(F_i)$. Associate each edge of F_i with a vertex of G as follows: choose an arbitrary vertex u of each tree T in forest F_i as the root of T; regard T as a rooted tree with root u in which all the edges are directed from the root to the descendants; and associate each edge e of tree T with the head vertex $h(e)$ of e. Thus, every vertex of F_i, except the roots, is associated with exactly one edge of F_i. Then we have

$$\sum_{(u,v) \in E} \min\{d(u), d(v)\} \leq \sum_{1 \leq i \leq a(G)} \sum_{e \in E(F_i)} d(h(e))$$

$$\leq \sum_{1 \leq i \leq a(G)} \sum_{v \in V} d(v)$$

$$= 2a(G)m. \qquad \text{Q.E.D.}$$

We remark here that only the concept of arboricity is used in the analysis of the running time of algorithms and that any of the algorithms requires neither finding $a(G)$ nor decomposing a graph into the minimum number of edge-disjoint forests.

8.3. Listing triangles

The triangle detection problem often arises in many combinatorial problems such as (1) the minimum cycle detection problem [IR78], (2) the approximate Hamiltonian walk problem in maximal planar graphs [NAW83] (see Section 10.4), and (3) the approximate minimum vertex cover (or maximum independent set) problem in planar graphs in [Alb74, BE82]. Itai and Rodeh [IR78] have presented an algorithm for finding all the triangles, which uses an adjacency matrix, so spends $O(n^2)$ space but runs in $O(m^{3/2})$ time for general graphs and in $O(n)$ time for planar graphs. Bar-Yehuda and Even [BE82] improved the space complexity of the algorithm from $O(n^2)$ into $O(n)$ by avoiding the use of the adjacency matrix. On the other hand Papadimitriou and Yannakakis [PY81] reported on a linear algorithm for finding all the cliques, i.e. K_i ($1 \leq i \leq 4$), in a planar graph.

The algorithm, given by Chiba and Nishizeki [CN85a], for listing triangles in a graph G is very simple as shown below. Observe that each triangle containing a vertex v corresponds to an edge joining two neighbours of v.

procedure K3(G);
    {Let G be a graph with n vertices and m edges.}
    begin
        sort the vertices $v_1, v_2, \ldots, v_n$ of G in such a way that $d(v_1) \geq d(v_2) \geq \cdots \geq d(v_n)$;
        for $i = 1$ *to* $n - 2$
            do begin
                {find all the triangles containing vertex v_i, each of which corresponds to an edge joining two neighbours of v_i.}
1:           mark all the vertices adjacent to v_i;
                for each marked vertex u
                    do begin
2:                *for* each vertex w adjacent to u
                    do if w is marked
                        then print out triangle (v_i, u, w);
3:                erase the mark from u
                end;
                {delete v_i from G so that no duplication occurs.}
4:           delete vertex v_i from G and let G be the resulting graph
        end
    end;

We have the following result on the algorithm.

Theorem 8.1. *Let G be a connected graph, then algorithm K3 lists all the triangles in G in $O(a(G)m)$ time, and especially in $O(n)$ time if G is planar.*

Proof. Since one can easily verify the correctness, we shall show that the algorithm runs in $O(a(G)m)$ time.

Clearly the degrees of vertices can be computed in $O(m)$ time. Since the degree of any vertex is at most $n - 1$, one can sort the vertices in $O(n)$ time by the bucket sort [AHU74]. Using adjacency lists, we can delete a vertex v from G in $O(d(v))$ time, and scan all the vertices adjacent to a vertex v in $O(d(v))$ time. Now consider the time required by the ith iteration of the outmost *for* statement. Statements 1, 3 and 4 spend $O(d(v_i))$ time. Statement 2 requires at most $O(\sum_{u \in N(v_i)} d(u))$ time, where $d(u)$ denotes the degree of vertex u in the original graph and $N(v_i)$ denotes the set of neighbours of v_i in the current graph. Therefore the total running time T of the algorithm is bounded as follows:

$$T \leqslant O(m) + O(n) + \sum_{v_i \in V} O\left(d(v_i) + \sum_{u \in N(v_i)} d(u)\right).$$

Since v_i has the largest $d(v_i)$ among all the vertices in the current graph, we have $d(u) \leqslant d(v_i)$ for each $u \in N(v_i)$. Since v_i is deleted at Statement 4, each edge of G is involved exactly once in the double summations above. Thus we have

$$T \leqslant O(m) + O(n) + O\left(\sum_{(u,v) \in E} \min\{d(u), d(v)\}\right).$$

Using Lemma 8.2, we have $T \leqslant O(a(G)m)$.

If G is planar, the algorithm runs in $O(a(G)m) \leqslant O(n)$ time since $a(G) \leqslant 3$ by Lemma 8.1(c).   Q.E.D.

Observe the following fact: algorithm K3 finds a triangle (K_3) containing a vertex v by detecting an edge (K_2) in a subgraph induced by the neighbors of v. In a similar manner, one can find a clique K_l containing a vertex v by detecting a clique K_{l-1} in a subgraph induced by the neighbors of v. An $O(la(G)^{l-2}m)$ time algorithm for listing all the cliques K_l in G based on this approach is found in [CN85a]. The algorithm lists all K_4 contained in a planar graph in linear time.

The benefit of the edge-searching strategy may be intuitively explained as follows: since we delete the vertices one by one in the largest degree order, the graph tends to quickly become sparse; this also prevents the edges incident on a vertex of large degree from being scanned very often.

Applying the strategy, we will give two more algorithms for other subgraph listing problems in the succeeding sections.

8.4. Listing quadrangles

In this section, we present an algorithm for finding all the quadrangles given in [CN85a].

If vertices $u_1, u_2, \ldots, u_l (l \geqslant 2)$ are all adjacent to two common vertices v and w, that is, these $l + 2$ vertices induce a complete bipartite graph $K_{2,l}$, then any quadruple (v, u_i, w, u_j), $1 \leqslant i \leqslant j \leqslant l$, forms a quadrangle. Thus even in a planar graph, there may exist $O(n^2)$ quadrangles. Instead of listing these quadrangles individually, we list a triple $(v, w, \{u_1, u_2, \ldots, u_l\})$ representing them altogether.

The algorithm C4 depicted below proceeds, for each vertex v of a graph, to find all the quadrangles containing v: for each vertex w within a distance two from v, the algorithm finds all such $u_1, u_2, \ldots, u_l$ which are adjacent to both v and w, and stores them in a set $U(w)$. When the quadrangles containing v have been found, v is deleted in order to avoid the duplication.

procedure C4(*G*);
  {Let $G = (V, E)$ be a graph with n vertices.}
  begin
    sort the vertices in V in a way that $d(v_1) \geqslant d(v_2) \geqslant \cdots \geqslant d(v_n)$;
    for each vertex $v \in V$ *do* $U(v) := \varnothing$;
    for $i = 1$ *to* n
      do begin
        for each vertex u adjacent to v_i
          do for each vertex $w \neq v_i$ adjacent to u
            do begin
              $U(w) := U(w) \cup \{u\}$
            end;
        for each vertex w with $|U(w)| \geqslant 2$
          do print out the triple $(v_i, w, U(w))$;
        for each vertex w with $U(w) \neq \varnothing$ *do* $U(w) := \varnothing$;
        delete the vertex v_i from G and let G be the new graph
      end
  end;

The graph depicted in Fig. 8.2 contains seven quadrangles. Algorithm C4 lists the following five triples: $(1, 5, \{2, 7, 10\})$, $(1, 4, \{2, 3\})$, $(3, 8, \{4, 6\})$, $(3, 9, \{4, 6\})$, and $(4, 6, \{8, 9\})$. The first triple represents three quadrangles.

We easily obtain the following theorem.

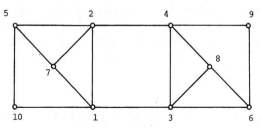

Fig. 8.2. A graph containing seven quadrangles.

Theorem 8.2. *Algorithm C4 obtains a representation of all the quadrangles in a connected graph G in* $O(a(G)m)$ *time, using* $O(m)$ *space.*

Note that Algorithm C4 does not store the triples. Since Algorithm C4 runs in $O(a(G)m)$ time, clearly all the quadrangles, if desired, could be represented by the triples in $O(a(G)m)$ spaces.

8.5. Listing maximal cliques

Tsukiyama et al. presented a backtracking algorithm which lists all the maximal independent sets in a graph G and requires at most $O(mn)$ time per maximal independent set [TIAS77]. By employing a similar idea together with the strategy, Chiba and Nishizeki have provided an algorithm which lists all the maximal cliques in a graph G in $O(a(G)m)$ time per maximal clique [CN85a]. In this chapter we present the prototype version of the algorithm which requires at most $O(mn)$ time per maximal clique.

Sometimes we must struggle with problems that appear to offer no efficient approach except for an exhaustive searching. In such situation, a systematic but exhaustive searching called *backtracking* will help us to find a solution. Backtracking finds a solution by continually trying to extend a partial solution. If it fails to extend a current partial solution then for other tries it "backtracks" to the most recent partial solution from which the current solution is extended. This process is often sketched by a *search tree*, an example of which will soon be presented.

The backtracking algorithm for listing all the maximal cliques is outlined as follows. Let $G = (V, E)$ be a given graph with vertex set $V = \{1, 2, \ldots, n\}$. Each vertex is referred by the number. Let G_i, $1 \leqslant i \leqslant n$, be the subgraph of G induced by vertices $1, 2, \ldots, i$. Assume that C is a maximal clique of G_{i-1}, then one can decide by the following lemma whether C or $(C \cap N(i)) \cup \{i\}$ is a maximal clique in G_i.

Lemma 8.3 [maximality test]. *Let C be a maximal clique of* G_{i-1}. *Then,*
(1) *C is a maximal clique of* G_i *if and only if* $C - N(i) \neq \varnothing$, *and*
(2) $(C \cap N(i)) \cup \{i\}$ *is a maximal clique of* G_i *if and only if* G_i *has no vertex* $y \in N(i) - C$ *such that* $y < i$ *and* $N(y) \supset C \cap N(i)$.

Proof. Immediate.   Q.E.D.

Hence we can generate all the maximal cliques of G_i from the maximal cliques of G_{i-1}. The following is the backtracking algorithm to generate all the maximal cliques of G. The set of vertices in a maximal clique C is also denoted by C.

procedure MAX-CLIQUE;
  procedure UPDATE(i, C);
    {generate a new maximal clique of G_i from a maximal clique C of G_{i-1}.}
    begin
      if $i = n + 1$
      then print out a new maximal clique C
      {C is a maximal clique of $G = G_n$.}
      else
      begin
        if $C - N(i) \neq \varnothing$ *then* UPDATE($i + 1$, C); {C is a maximal clique of G_i.}
        if there is no vertex $y \in N(i) - C$ such that $y < i$ and $N(y) \supset C \cap N(i)$
        then
          begin
            SAVE := $C - N(i)$; {save the vertices removed from current C}
            $C := (C \cap N(i)) \cup \{i\}$; {new C is a maximal clique of G_i.}
            UPDATE($i + 1$, C);
            $C := (C - \{i\}) \cup$ SAVE {recovery to old C}
          end
      end
    end;

  begin
    $C := \{1\}$; {C is the unique maximal clique of G_1.}
    UPDATE(2, C)
  end;

In Fig. 8.3 we illustrate an input graph and the search tree. Graphs G_i, $i \geqslant 2$, which appear in the searching are drawn at nodes of the search tree. Every G_{i-1} is circled and the vertices in C are drawn as black points. For each node G_{i-1}, its left son G_i is drawn if C passes test (1) in Lemma 8.3, while its right son G_i is drawn if $(C \cap N(i)) \cup \{i\}$ passes test (2).

Duplications may occur in maximal cliques produced by rule (2). For example in Fig. 8.3 maximal cliques $\{3, 4, 5\}$ and $\{3, 5, 6\}$ are generated twice; they stem from the two maximal cliques $\{3, 4\}$ of G_4 drawn in double circles; the left one is generated from the maximal clique $\{1, 3\}$ of G_3, and the right one from $\{2, 3\}$, both by rule (2). However the duplications can be avoided if one

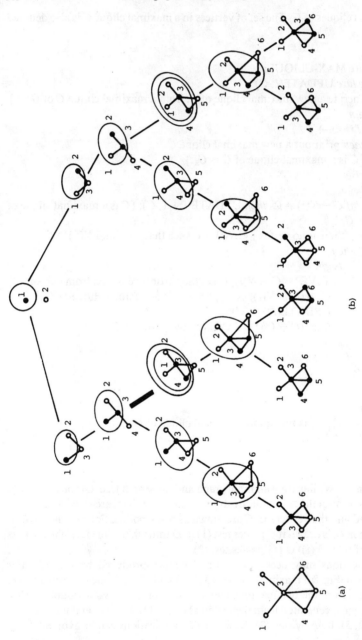

Fig. 8.3. (a) An input graph *G* and (b) a search tree.

chooses the lexicographically largest one among all the cliques C having the same $C \cap N(i)$. In the example above $\{3, 4\}$ of G_4 is generated only from $\{2, 3\}$ of G_3 since $\{2, 3\}$ is the lexicographically largest among all maximal cliques C of G_3 having the same $C \cap N(4) = \{3\}$.

Testing the lexicographically largest can be performed as follows.

Lemma 8.4 [lexico. test]. *Let C be a maximal clique of G_{i-1} and let $C_0 = C \cap N(i)$. Then a maximal clique C of G_i is the lexicographically largest one among all the maximal cliques containing C_0 if and only if there is no vertex $y \notin C$ such that $N(y) \supset C_0 \cup C^y$, where $C^y = \{k \in C \mid k > y\}$.*

Proof. *Necessity*: Assume that there exists a vertex $y \notin C$ such that $N(y) \supset C_0 \cup C^y$. Then clearly there exists a maximal clique containing $\{y\} \cup C_0 \cup C^y$ which is lexicographically larger than C.

Sufficiency: Assume that there exists a maximal clique C' ($\supset C_0$) which is lexicographically larger than C. Let y be the largest vertex in $C' - C$. Then $C \cap C' \supset C^y$ since every vertex in $(C - C')$ is less than y. Thus we have $N(y) \supset C \cap C' \supset C_0 \cup C^y$.   Q.E.D.

We add this lexico. test into the third *if*-statement in MAX-CLIQUE. If the modified algorithm is applied to G in Fig. 8.3(a), then the edge drawn in a thick line can be "pruned" in the search tree. We now have the following theorem on modified MAX-CLIQUE.

Theorem 8.3. *Algorithm MAX-CLIQUE lists all the maximal cliques of a connected graph G in $O(mn)$ time per maximal clique, using $O(m)$ space.*

Proof. Using Lemmas 8.3 and 8.4, one can prove the correctness. Therefore we shall concentrate on the claim on time and space.

Consider the time spent by UPDATE(i, C), excluding the time required by the recursive calls. It is an easy exercise to show that both "maximality test" and "lexico. test" can be performed once in $O(m)$ time. Furthermore one can easily observe that bookkeeping operations for saving and recovery of C spend at most $O(n)$ time. Hence the total time required to generate one maximal clique is at most $O(mn)$.

Every UPDATE(i, C), $i \leqslant n$, calls at least once UPDATE$(i + 1, C)$. In fact, if the first recursive call in UPDATE(i, C) does not occur, then the second necessarily occurs. Thus every call of UPDATE eventually generates at least one maximal clique, and hence the time spent by any statement is counted in the time above at least once for some maximal clique of G. Thus we have shown that MAX-CLIQUE requires $O(mn)$ time per maximal clique.

Since set C is a global variable, C requires $O(n)$ space. Since the sets of vertices contained in the local variable SAVE are pairwise disjoint, SAVE requires $O(n)$ space in total. The adjacency lists require at most $O(m)$ space. Thus MAX-CLIQUE spends at most $O(m)$ space in total.   Q.E.D.

More sophisticated implementation of maximality and lexico. tests improves the bound to $O(a(G)m)$ [CN85a].

CHAPTER 9

PLANAR SEPARATOR THEOREM

9.1. Introduction

One efficient approach to solving computational problems for graphs is "divide-and-conquer." In this method, the original graph is divided into two or more smaller graphs. The problems for subgraphs are solved by applying the method recursively, and the solutions to the subgraphs are combined to give the solution to the original problem. The planar separator theorem of Lipton and Tarjan provides a basis for the approach [LT79]. The theorem asserts that any planar graph of n vertices can be divided into components of roughly equal size by removing only $O(n^{1/2})$ vertices.

Very recently Miller generalized the planar separator theorem [Mil84]. His theorem gives us a separator which is a single cycle, that is, a "cycle separator." Algorithms for some applications can be simplified by the cycle separator [JV82, Ric86].

In Section 9.2 we define terms and present some results which can be easily observed. In Section 9.3 we present Miller's proof. In the last two sections we present some applications of a planar separator theorem.

9.2. Preliminary

Formally we say a class of graphs has an $f(n)$-*separator* if the following statement is true:

For any graph G of n vertices in the class, there exist constants $a < 1, b > 0$ such that the vertices of G can be partitioned into three sets A, B, C such that no edge joins a vertex in A with a vertex in B, neither A nor B contains more than an vertices and C contains no more than $bf(n)$ vertices.

As will be shown in Section 9.4, many problems concerning graphs in the class can be solved efficiently using the divide-and-conquer method if the vertex set C can be found efficiently.

As illustrations, we present the following $f(n)$-separator theorems which can be rather easily verified.

Theorem 9.1. *The class of trees has a 1-separator.*

Proof. We first show that, by removing a single vertex, one can partition any tree T into two or more subtrees having no more than $2n/3$ vertices.

Pick an arbitrary vertex of T as the root r. If every son of r has, including itself, no more than $2n/3$ descendants, then r is a desired vertex. Otherwise, let c be the son of r having more than $2n/3$ descendants, and iterate a similar test towards leaves by redefining c until no son of c has more than $2n/3$ descendants. Clearly such c necessarily exists. Partition T into subtrees by removing c from T. The subtree which contains root r has no more than $n/3$ vertices, since the subtree of T rooted by c has more than $2n/3$ vertices. Clearly other subtrees produced by removing c have no more than $2n/3$ vertices.

Now one can easily find a desired partition of T into two sets of subtrees A and B each of which has no more than $2n/3$ vertices as follows. If there exists a subtree having more than $n/3$ vertices, then let A be the subtree and let B be the set of remaining subtrees. Otherwise, that is, if every subtree has no more than $n/3$ vertices, then let A be a maximal subset of subtrees which together contain no more than $2n/3$ vertices, and let B be the set of remaining subtrees. In either case B contains no more than $2n/3$ vertices. Hence $C = \{c\}$ is a desired separator.   Q.E.D.

An *outerplanar* graph is a planar graph which has an embedding such that all its vertices lie on the same face, say the outer face. A 2-connected outerplanar graph G is a triangulation of n-gon, perhaps with some edges removed. Delete from the dual G^* of G the vertex corresponding to the outer face of G. The resulting graph is a tree. Noting this fact, one can easily verify the following theorem in the same way as for Theorem 9.1.

Theorem 9.2. *[Lei 80].* *The class of 2-connected outerplanar graphs has a 1-separator.*

Proof. Embed the given outerplanar graph so that all the vertices lie on the outer face, and then triangulate all faces except the outer face. Thus every facial cycle except the outer facial cycle is a triangle. It suffices to show that there exists a facial cycle whose removal produces (at most three) connected components having no more than $2n/3$ vertices, since clearly such a facial cycle is a desired separator.

Pick an arbitrary facial cycle uvw such that the vertices u, v and w appear on the outer facial cycle in clockwise order, and let P_{uv}, P_{vw} and P_{wu} be the paths running through the outer facial cycle clockwise from u, v and w to v, w and u, respectively. If the removal of u, v, w produces no connected component having more than $2n/3$ vertices, then the cycle uvw is a desired facial cycle. Otherwise, let $P_{uv} - \{u, v\}$ has more than $2n/3$ vertices without loss of generality; pick the other facial cycle having an edge (u, v); and do the same operation. Note that $P_{vw} \cup P_{wu} - \{u, v\}$ has at most $n/3$ vertices since $P_{uv} - \{u, v\}$ has at least $2n/3$ vertices. Thus, by iterating such an operation, we can eventually find a desired facial cycle. Q.E.D.

Theorem 9.3. *The class of $n^{1/2} \times n^{1/2}$ grid graphs (meshes) has an $n^{1/2}$-separator.*

Proof. One can easily verify the correctness of the claim. Q.E.D.

Trees, outerplanar graphs and grid graphs are subsumed by planar graphs. Lipton and Tarjan proved that the class of planar graphs has an $n^{1/2}$-separator [LT79]. Their planar separator theorem is formally described as follows.

Theorem 9.4. *[Lipton and Tarjan]. Let G be any planar graph with n vertices having non-negative vertex costs summing to no more than one. Then the vertices of G can be partitioned into three sets A, B, C such that no edge joins a vertex in A with a vertex in B, neither A nor B has total cost exceeding $\frac{2}{3}$, and C contains no more than $2(2n)^{1/2}$ vertices.*

In the succeeding section we prove Miller's cycle separator theorem, which implies the existence of a $n^{1/2}$-separator in planar graphs.

9.3. Planar separator theorem

Let G be a 2-connected plane graph with weights on faces and vertices which sum to 1. (See Fig. 9.1.) For a (simple) cycle C of G we denote by $w(C)$ the sum of weights of vertices on C and call $w(C)$ the *total weight of cycle C*. Further we denote by int(C) the interior of C, and by $w(\text{int}(C))$ the sum of the weights of faces and vertices in int(C). Similarly we define the exterior ext(C) and the weight $w(\text{ext}(C))$. A cycle of G is a *weighted separator* if $w(\text{int}(C)) \leq \frac{2}{3}$ and $w(\text{ext}(C)) \leq \frac{2}{3}$. In Fig. 9.1 a weighted cycle separator C of G is drawn by thick lines. In this chapter we denote by z the *maximum face size* of G (i.e. the maximum number of edges contained in a facial cycle of G), and hence $z = 3$ if G is maximal planar. We shall not distinguish between a face and its boundary. Miller has given the following theorem on a weighted separator.

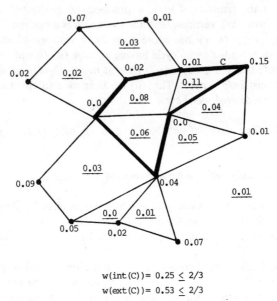

$$w(\text{int}(C)) = 0.25 \le 2/3$$
$$w(\text{ext}(C)) = 0.53 \le 2/3$$

Fig. 9.1. A plane graph with weights on vertices and faces.

Theorem 9.5. [*Mil*84]. *Let G be a 2-connected plane graph with weights on faces and vertices which sum to 1, and no face having weight $> \frac{2}{3}$. Then there exists a weighted cycle separator of size at most $4(2zn)^{1/2}$, and the cycle can be found in linear time.*

Theorem 9.5 immediately implies the existence of a separator of size at most $4(6n)^{1/2}$, for the class of planar graphs, not necessarily 2-connected. Note that the maximal planar graph augmented from an arbitrary planar graph G by adding extra edges has the maximal face size $z = 3$.

In the remainder of this section we prove Theorem 9.5. If either a face has a weight $\ge \frac{1}{3}$ or the total weight of a facial cycle is $\ge \frac{1}{3}$, then the facial cycle is a weighted cycle separator of G. Therefore we may assume in the following arguments that the weight of every face and the total weight of every facial cycle are $< \frac{1}{3}$.

The proof of Theorem 9.5 will be completed by two lemmas. The first lemma claims that any 2-connected plane graph G necessarily contains a subgraph H with a spanning tree T such that $D + z_H \leqslant 4(2zn)^{1/2}$ where D is the *diameter* of T (i.e. the number of edges in a longest path in T) and z_H is the *maximum face size* of H. The second lemma claims that such a graph H whose faces F have "induced weights" $w(\text{int}(F))$ contains a weighted cycle separator of size at most $D + z_H$, which naturally induces a desired cycle separator of G. It should be noted that all the procedures which will appear in the proof of lemmas can be performed in linear time.

We define some terms before proving the first lemma. If C is a cycle of a plane graph G, and if C' is the ring sum of C and the faces in $\text{ext}(C)$ which share a vertex or an edge with C, then C' is either empty or a disjoint union of cycles; each of the cycles is called a *front* of C. For example in Fig. 9.2 the fronts of cycle 1 are cycles 2, 3, 4 and 5. The interior of a front is defined to be the side containing C. Starting with a facial cycle F, construct the fronts of F, their fronts, and so on. The collection of all these fronts is called *the fronts of a face F*. In Fig. 9.2 the fronts of a face F are drawn by thick lines. The relation "C' is a front of C" defines a tree T_F, called a *front tree*, whose nodes correspond to the fronts and edges represent the relation. A leaf of the tree corresponds either to a single face or to an empty front. An empty front has a total weight 0 and a size 0. Fig. 9.2(b) illustrates a front tree, whose leaves 11, 12 and 13 represent empty fronts. Let H be a plane subgraph of G and let F be a face of H, then the *induced weight* of face F is the weight $w(\text{int}(F))$ for cycle F in G.

Lemma 9.1. *Let G be a 2-connected plane graph with weights on its faces and vertices summing to 1, and no face having weight $> \frac{2}{3}$, and let z be the maximum face size. Then there exists a 2-connected subgraph H with a spanning tree T such that:*

(1) *the diameter D of T and the maximum face size z_H of H satisfies*

$$D + z_H \leqslant 4(2zn)^{1/2},$$

(2) *every face F of H has an induced weight $w(\text{int}(F)) \leqslant \frac{2}{3}$.*

Proof. Choose any face F of G, and let T_F be the front tree. (See Fig. 9.2.) It is an easy exercise to compute the weights $w(\text{int}(C))$ and $w(\text{ext}(C))$ for all the fronts C in linear time. Let B be a "maximal" front such that $w(\text{int}(B)) < \frac{1}{3}$ and

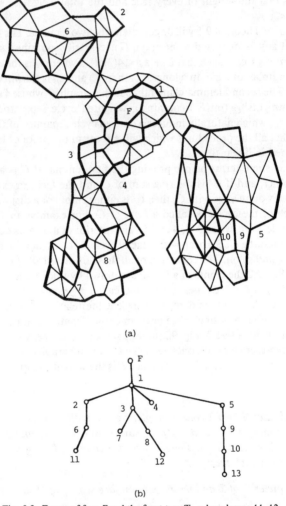

(a)

(b)

Fig. 9.2. Fronts of face F and the front tree T_F, where leaves 11, 12, and 13 are empty fronts.

B has no descendant B' with $w(\text{int}(B')) < \frac{1}{3}$. Remember that $\text{int}(B)$ is the side containing F. Every leaf front L satisfies $w(\text{int}(L)) > \frac{1}{3}$; because, if L is an empty front then $w(\text{int}(L)) = 1$, otherwise L is a single face and hence the weight of face L plus the total weight of the facial cycle is $< \frac{2}{3}$. Thus B must not be a leaf front. Let $B = B_0, B_1, \ldots, B_k = F$ be the ancestors of B up to F in T_F. We now claim that one of them, say B_{l_1}, satisfies

$$zl_1 + \text{SIZE}(B_{l_1}) \leqslant (2zn)^{1/2}. \tag{9.1}$$

Assume on the contrary that $\text{SIZE}(B_i) > (2zn)^{1/2} - zi$ for every i, $0 \leqslant i \leqslant k$. Then $z \geqslant \text{SIZE}(B_k = F) > (2zn)^{1/2} - zk$, and hence $k > (2n/z)^{1/2} - 1$. Furthermore we have

$$\sum_{0 \leqslant i \leqslant k} \text{SIZE}(B_i) > \sum_{0 \leqslant i \leqslant k} ((2zn)^{1/2} - zi).$$

However, since B_i's are disjoint, the left hand side is $\leqslant n$, while by a straightforward calculation the right hand side is $\geqslant n$. This is a contradiction. Thus B_{l_1} necessarily exists.

Construct a subgraph H' from G by deleting

(1) the exterior of all fronts of B_i, $0 \leqslant i \leqslant l_1$, which are distinct from B_j themselves, $0 \leqslant j \leqslant l_1$; and

(2) the interior of B_{l_1}.

In Fig. 9.3 deleted parts are hatched. Since $w(\text{int}(B_0)) < \frac{1}{3}$ and all the fronts ($\neq B_0$) of B_i, $1 \leqslant i \leqslant l_1$, lie in the interior of B_0, the new faces produced by the deletions have induced weights $< \frac{1}{3}$. Moreover the exterior of any front of B_0 has a weight $\leqslant \frac{2}{3}$ since B_0 is a maximal front of $w(\text{int}(B_0)) < \frac{1}{3}$. Thus every face of H' has an induced weight $\leqslant \frac{2}{3}$. However sizes of new faces may be too large. So we modify H' as follows. Let H'' be the graph obtained from H' by recovering $\text{ext}(C)$ for every face C ($\neq B_{l_1}$) of size $> z$. If $C = C_0, C_1, \ldots, C_r$ are all the descendants on a path of T_F going from C to a leaf C_r, then by similar arguments as used above, one can easily show that there exists a front C_{l_2}, $1 \leqslant l_2 \leqslant r$, such that

$$zl_2 + \text{SIZE}(C_{l_2}) \leqslant (2zn)^{1/2}. \tag{9.2}$$

Delete the exterior of front C_{l_2} for each of the leaf descendants of C. Repeat the above operation for all C ($\neq B_{l_1}$) of size $> z$. H is the resultant graph. Clearly any face of H has an induced weight $\leqslant \frac{2}{3}$. Therefore we shall show that a spanning tree T of H satisfies the condition (1) in Lemma 9.1.

If a front R' is a son of a front R and x is a vertex on R', then the distance from x to R is at most $z/2$ because x is on a face incident to R whose size is at most z. Therefore the distance from a vertex of H to B_{l_1} is at most $(z(l_1 + 1) + zl_2)/2$. Thus the distance from any vertex to a particular vertex v of B_{l_1} is at most $(z(l_1 + 1) + zl_2 + \text{SIZE}(B_{l_1}))/2$. Therefore the breadth-first search starting from v defines a spanning tree T of diameter D such that

$$D \leqslant z(l_1 + 1) + zl_2 + \text{SIZE}(B_{l_1}). \tag{9.3}$$

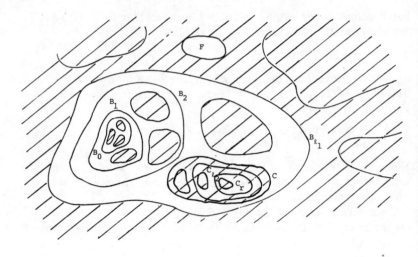

(a) Fronts

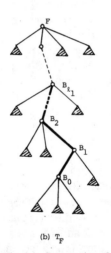

(b) T_F

Fig. 9.3. Construction of H'.

Clearly the maximum face size z_H of H satisfies

$$z_H \leqslant \max\{z, \text{SIZE}(B_{l_1}), \text{SIZE}(C_{l_2})\}. \tag{9.4}$$

From (9.1)–(9.4) we have

$$D + z_H \leqslant z(l_1 + 1) + zl_2 + \text{SIZE}(B_{l_1}) + \max\{z, \text{SIZE}(B_{l_1}), \text{SIZE}(C_{l_2})\}$$

$$\leqslant zl_1 + \text{SIZE}(B_{l_1}) + zl_2 + \text{SIZE}(C_{l_2}) + z + \max\{z, \text{SIZE}(B_{l_1})\}$$

$$\leqslant \max\{zl_1 + \text{SIZE}(B_{l_1}) + zl_2 + \text{SIZE}(C_{l_2}) + 2z,$$

$$2(zl_1 + \text{SIZE}(B_{l_1})) + zl_2 + \text{SIZE}(C_{l_2}) + z\}$$

$$< 4(2zn)^{1/2}. \qquad\qquad \text{Q.E.D.}$$

Before presenting the second lemma, we introduce some notations. Refer to a 2-connected plane graph G with a spanning tree T illustrated in Fig. 9.4(a), where only edges concerning with the following arguments are drawn, tree edges by solid lines and non-tree edges by wavy lines. Let $e = (x, y)$ be any non-tree edge and let C_e be the unique cycle in $T + e$. Assume that $w(\text{int}(C_e)) > \frac{2}{3}$ and any single face has a weight $< \frac{1}{3}$. Let F be the facial cycle that contains e and lies in the interior of C_e. Let $e_1, e_2, \ldots, e_k$ be the non-tree edges on F distinct from e, and let C_i be the unique cycle in $T + e_i$. Then $k \geqslant 1$ since any single face has a weight $< \frac{1}{3}$. Let $C_f = \Sigma_{1 < i < k} C_i$ where the sum operation denotes ring sum $\oplus$. Let x and y be the ends of the edge e. Remove e from F, then we obtain a path P_{xy} from x to y on F. Let $x = x_1, x_2, \ldots, x_t = y$ be the vertices on P_{xy} in the order they appear. Any cycle C_i has a leftmost vertex x_{i_l} (i.e. one with the minimum index i_l) and a rightmost vertex x_{i_r} (i.e. one with the maximum index i_r) in $\{x_1, x_2, \ldots, x_t\}$. For two distinct cycles C_i and C_j among $C_f, C_1, \ldots, C_k$ we say C_i *domains* C_j if $i_l \leqslant j_l \leqslant j_r \leqslant i_r$. Further, we say C_i *directly domains* C_j if there is no C_l such that C_i domains C_l and C_l domains C_j. For example, in Fig. 9.4, C_{13} domains C_{14}, C_{15}, C_{16} and C_{17}, but does not directly domain C_{14}. Clearly this relation "C_i domain C_j" is a partial order, and we can define a *domain tree* having nodes C_i and edges representing the relation "C_i directly domain C_j", which is illustrated in Fig. 9.4(b). Let $C_i^* = C_i \oplus \Sigma\{C_j \mid C_i \text{ domains } C_j\}$, then C_i^* is necessarily a single cycle. Let $C_{i_1}, C_{i_2}, \ldots, C_{i_s}$ be the sons of C_i listed in the order the first half of the sequence of intersections $P_{xy} \cap C_{i_j}^*, j = 1, 2, \ldots, s$, successively appear on P_{xy} from x to e_i and the latter half from y to e_i. Let $D_j = C_{i_j}^*, j = 1, 2, \ldots, s$. In Fig. 9.4(a), for example, if $C_i = C_8$ then $D_1 = C_2^*$, $D_2 = C_3^*$, $D_3 = C_5^*$, $D_4 = C_6^*$, $D_5 = C_7^*$, $D_6 = C_{11}^*$ and $D_7 = C_9^*$. Although ring sum $F \oplus D_i \oplus D_j$ is not necessarily a single cycle for arbitrary indices i and j (for example $F \oplus C_{13}^* \oplus C_{22}^*$ in Fig. 9.4), ring sum of form $F \oplus D_1 \oplus \cdots \oplus D_j, 1 \leqslant j \leqslant s$, is necessarily a single cycle.

We are now ready to present and prove Lemma 9.2, which together with Lemma 9.1 immediately implies Theorem 9.5.

Lemma 9.2. *Let G be a 2-connected plane graph with a spanning tree T and no*

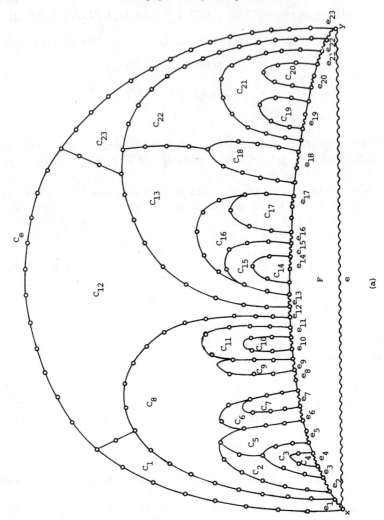

Fig. 9.4. (a) Cycle C_e and cycles domained by C_f; (b) Domain tree.

face having weight $> \frac{2}{3}$. Then there exists a weighted cycle separator of size at most $D + z$, where D is the diameter of T and z the maximum face size of G.

Proof. We may assume that each face has weight $< \frac{1}{3}$; otherwise, the facial cycle would be a desired weighted cycle separator. Choose an arbitrary non-

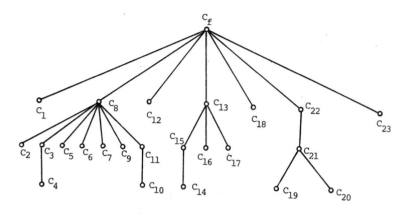

(b)

Fig. 9.4 (continued)

tree edge e. If C_e is a weighted cycle separator, then C_e has size $\leq D + 1$ and hence the proof is completed. So assume without loss of generality that $w(\text{int}(C_e)) > \frac{2}{3}$. Repeat to reset e to e_i while there exists a cycle C_i such that $w(\text{int}(C_i)) > \frac{2}{3}$. Thus we may assume that C_i satisfies $w(\text{int}(C_i)) \leq \frac{2}{3}$ for each non-tree edge e_i, $1 \leq i \leq k$, on F. If $w(\text{ext}(C_i)) \leq \frac{2}{3}$ for some $1 \leq i \leq k$, then C_i is a weighted cycle separator of size $\leq D + 1$. If $w(\text{ext}(F)) \leq \frac{2}{3}$, then F is a weighted cycle separator of size $\leq z$. Thus the problem is reduced to the case when all weights $w(\text{int}(C_e))$, $w(\text{ext}(F))$, and $w(\text{ext}(C_i))$, $1 \leq i \leq k$, are greater than $\frac{2}{3}$.

We now claim that either

(a) there exists a weighted cycle separator $F \oplus C_i^*$, $i = 1, 2, \ldots,$ or k; or

(b) the domain tree has a non-leaf node C_i such that $w(\text{int}(F \oplus C_i^*)) > \frac{2}{3}$ and $w(\text{ext}(F \oplus D_j)) > \frac{2}{3}$ for every son C_{i_j} of C_i.

Suppose that no weighted cycle separator $F \oplus C_i^*$ exists. Then either $w(\text{int}(F \oplus C_i^*)) > \frac{2}{3}$ or $w(\text{ext}(F \oplus C_i^*)) > \frac{2}{3}$. Since $w(\text{ext}(C_i)) > \frac{2}{3}$, $w(C_i) + w(\text{int}(C_i)) < \frac{1}{3}$. If C_i is a leaf of the domain tree, then $C_i^* = C_i$ and $w(\text{int}(F \oplus C_i)) \leq w(\text{int}(F)) + w(C_i) + w(\text{int}(C_i)) < \frac{2}{3}$. Let C_i be a minimal cycle in a sense that $w(\text{int}(F \oplus C_i^*)) > \frac{2}{3}$ and every son C_{i_j} of C_i satisfies $w(\text{int}(F \oplus D_j)) \leq \frac{2}{3}$. We can necessarily choose such a cycle C_i among the non-leaf nodes of the domain tree since $w(\text{int}(F)) < \frac{1}{3}$, $w(\text{int}(C_e)) > \frac{2}{3}$ and $C_e = F \oplus C_f^*$. Each son C_{i_j} of C_i satisfies $w(\text{ext}(F \oplus D_j)) > \frac{2}{3}$; otherwise $F \oplus D_j$ would be a weighted cycle separator. Thus we have shown that either (a) or (b) above holds.

Since $F \oplus C_i^*$ has size at most $D + z$, we may assume that (b) above holds. Then we claim that there exists j, $2 \leqslant j \leqslant s$, such that $F \oplus D_1 \oplus \cdots \oplus D_j$ is a weighted cycle separator. Since $w(\text{ext}(C_i)) > \frac{2}{3}$, $w(\text{int}(C_i)) < \frac{1}{3}$. Further we have $w(\text{ext}(F \oplus D_1 \oplus \cdots \oplus D_s)) \leqslant \frac{2}{3}$; otherwise,

$$w(\text{int}(F \oplus D_1 \oplus \cdots \oplus D_s)) + w(F \oplus D_1 \oplus \cdots \oplus D_s) < \tfrac{1}{3}$$

and hence

$$w(\text{int}(F \oplus C_i^*)) \leqslant w(\text{int}(C_i)) + w(\text{int}(F \oplus D_1 \oplus \cdots \oplus D_s))$$
$$+ w(F \oplus D_1 \oplus \cdots \oplus D_s)$$
$$< \tfrac{2}{3},$$

contradicting (b) above. Thus $w(\text{ext}(F \oplus D_1 \oplus \cdots \oplus D_s)) \leqslant \frac{2}{3}$ and $w(\text{ext}(F \oplus D_1)) > \frac{2}{3}$. Therefore there exists an index j, $2 \leqslant j \leqslant s$, such that $w(\text{ext}(F \oplus D_1 \oplus \cdots \oplus D_{j-1})) > \frac{2}{3}$ and $w(\text{ext}(F \oplus D_1 \oplus \cdots \oplus D_j)) \leqslant \frac{2}{3}$. Since $w(\text{ext}(F \oplus D_1 \oplus \cdots \oplus D_{j-1})) > \frac{2}{3}$,

$$w(\text{int}(F \oplus D_1 \oplus \cdots \oplus D_{j-1})) + w(F \oplus D_1 \oplus \cdots \oplus D_{j-1}) < \tfrac{1}{3}.$$

Since $w(\text{ext}(F \oplus D_j)) > \frac{2}{3}$,

$$w(\text{int}(D_j)) \leqslant w(\text{int}(F \oplus D_j)) < \tfrac{1}{3}.$$

Thus

$$w(\text{int}(F \oplus D_1 \oplus \cdots \oplus D_j))$$
$$\leqslant w(\text{int}(F \oplus D_1 \oplus \cdots \oplus D_{j-1})) + w(F \oplus D_1 \oplus \cdots \oplus D_{j-1})$$
$$+ w(\text{int}(D_j))$$
$$< \tfrac{2}{3}.$$

Hence we can conclude that $F \oplus D_1 \oplus \cdots \oplus D_j$ is a weighted cycle separator.

Finally we must show that the weighted cycle separator $F \oplus D_1 \oplus \cdots \oplus D_j$ has a desired size. The cycle consists of at most two paths on cycle F and at most two paths on T. Further we can easily observe that the paths on T are contained in the path $C_i - e_i$ of T. Thus the size of the separator is at most $D + z_H$.   Q.E.D.

9.4. Applications of the planar separator theorem

As an application of a separator theorem, Lipton and Tarjan have presented an approximation algorithm for the maximum independent set problem. The algorithm runs in O($n \log n$) time. The worst-case ratio, that is, the smallest ratio of the size of an approximate-solution to the size of a maximum solution taken over all problem instances is $1 - O(1/(\log \log n)^{1/2})$, asymptotically tending to 1 as $n \to \infty$ [LT80].

In this section we study the general conditions under which the Lipton and Tarjan's approach is applicable to combinatorial problems on planar graphs. A number of combinatorial problems are formulated as a "maximum induced subgraph problem" P with respect to some graph property Q; the problem asks for a maximum subset S of V that induces a subgraph satisfying Q. For example, the maximum independent set problem is a maximum induced subgraph problem P with property "independent (i.e. pairwise nonadjacent)." However it has been shown in a unified way that the maximum induced subgraph problem is NP-complete for various properties Q [KD79, Yan78]. In this section we first discuss the requirements for their approach to yield an efficient approximation algorithm for the maximum induced subgraph problem on planar graphs, and then give efficient approximation algorithms for the problems.

Lipton and Tarjan's algorithm for the maximum independent set problem uses the following form of their separator theorem.

Theorem 9.6. [*LT*80]. *Every planar graph of n vertices contains a set C of* O($(n/\varepsilon)^{1/2}$) *vertices whose removal leaves no connected component with more than εn vertices, where ε is any constant such that* $0 < \varepsilon < 1$. *Furthermore the set C can be found in* O($n \log n$) *time.*

Proof. The set C can be found as follows. First give the equal-costs $1/n$ to all the vertices and set $C = \varnothing$. Next repeat the following partition steps until $G - C$ has no component of more than εn vertices:
(1) find some connected component K having more than εn vertices, say exactly k vertices;
(2) reassign equal vertex-costs $1/k$ to all vertices of K;
(3) apply Theorem 9.4 to K to partition the vertex set of K into K_A, K_B and K_C; and
(4) let $C := C \cup K_C$.
Clearly the procedures above terminate finitely. Thus we shall show that the size of C is O($(n/\varepsilon)^{1/2}$) and the running time is O($n \log n$).

Consider all the connected components which appear during an execution of the above procedure. Assign a *level* to each component as follows: the final components have level 0; and each of the intermediate components has a level one greater than the maximum level of the components produced from it. Then obviously any two components of the same level are disjoint.

Since a component of level i has at least $(\frac{3}{2})^i$ vertices, the maximum level l must satisfy $(\frac{3}{2})^l \leqslant n$ or $l \leqslant \log_{3/2} n$. Since every component of level 1 has at least εn vertices, every component of level i has at least $(\frac{3}{2})^{i-1}\varepsilon n$ vertices. Therefore the number c_i of components of level i is at most $(\frac{2}{3})^{i-1}/\varepsilon$.

Now we can bound the size of C as follows. Let n_j^i, $1 \leqslant j \leqslant c_i$, be the number of vertices in the jth component of level i. Then we have

$$|C| \leqslant \sum_{1 \leqslant i \leqslant l} \sum_{1 \leqslant j \leqslant c_i} 2(2n_j^i)^{1/2}$$

$$\leqslant 2(2)^{1/2} \sum_{1 \leqslant i \leqslant l} \left(c_i \sum_{1 \leqslant j \leqslant c_i} n_j^i \right)^{1/2}$$

$$\leqslant 2(2)^{1/2} \sum_{1 \leqslant i \leqslant l} c_i^{1/2} n^{1/2}$$

$$\leqslant 2(2)^{1/2}(n/\varepsilon)^{1/2} \sum_{1 \leqslant i \leqslant l} (\tfrac{2}{3})^{(i-1)/2}$$

$$\leqslant O((n/\varepsilon)^{1/2}).$$

The total running time is bounded by

$$\sum_{1 \leqslant i \leqslant l} \sum_{1 \leqslant j \leqslant c_i} O(n_j^i) \leqslant \sum_{1 \leqslant i \leqslant l} O(n) \leqslant O(n \log n). \qquad \text{Q.E.D.}$$

From Lipton and Tarjan's approximation algorithm for the maximum independent set problem, we naturally have the following "algorithm" MISP for the maximum induced subgraph problem P with Q.

Algorithm MISP(ε).
Comment $0 < \varepsilon < 1$.

Step 1. Applying Theorem 9.6 to a given planar graph $G = (V, E)$, find a set of vertices C of size $O((n/\varepsilon)^{1/2})$, such that each connected component of $G - C$ contains at most εn vertices.

Step 2. In each connected component $G_i = (V_i, E_i)$ of $G - C$, find a maximum set S_i inducing a subgraph with property Q by checking exhaustively every subset of V_i.

Step 3. Form the approximate solution S as a union of maximum sets, one from each component, that is, $S = \sum S_i$.

One can easily observe that the conditions given in the lemma below are sufficient for the success of the algorithm MISP.

Lemma 9.3. *The approximation algorithm MISP(ε) has time complexity* $O(\max\{n \log n, n2^{\varepsilon n}\})$ *and worst-case ratio* $1 - O(1/(\varepsilon n)^{1/2})$ *if the following conditions are satisfied:*

(C1) *The subgraph of G induced by S satisfies property Q.*

(C2) *The error* $|S^*| - |S|$ *is bounded by the number of vertices of C, that is,* $|S^*| - |S| = O(|C|)$, *where S^* is a maximum vertex set inducing a subgraph of G with Q.*

(C3) $|S^*|$ *is a positive fraction of n.*

(C4) *Property Q is recognizable in linear time (that is, one can determine in linear time whether a graph satisfies Q or not).*

Proof. The subgraph of G induced by S is necessarily a solution of P by the condition (C1). The conditions (C2) and (C3) imply that the worst-case ratio is $|S|/|S^*| = 1 - (|S^*| - |S|)/|S^*| = 1 - O(1/(\varepsilon n)^{1/2})$. Step 1 requires at most $O(n \log n)$ time. In Step 2, $2^{|V_i|}$ subsets of V_i are checked, so the condition (C4) implies that this step can be done within $O(\sum |V_i| 2^{|V_i|}) \leqslant O(n2^{\varepsilon n})$ time. Thus the total running time of MISP is $O(\max\{n \log n, n2^{\varepsilon n}\})$.   Q.E.D.

We now introduce two more terms to specify a class of properties Q for which conditions (C1)–(C4) hold. A graph property Q is *hereditary* if every subgraph of G satisfies Q whenever G satisfies Q. Property Q is *determined by the components* if a graph G satisfies Q whenever every connected component of G satisfies Q. For example, the property "planar" is hereditary and determined by the components since every subgraph of a planar graph is planar and a graph is planar if and only if every connected component is planar. It has been known that a number of graph properties are hereditary and determined by the components [Yan78]. In order to avoid a trivial case, we now assume that at least one nonempty graph satisfies Q.

We have the following theorem from Lemma 9.3.

Theorem 9.7. [CNS81a]. *If P is a maximum induced subgraph problem with respect to property Q which is*

(a) *hereditary;*

(b) *determined by the components; and*

(c) *recognizable in linear time,*

then there exists an approximation algorithm for P on planar graphs with time complexity $O(n \log n)$ *and worst-case ratio* $1 - O(1/(\log \log n)^{1/2})$.

Proof. We claim that the conditions (C1)–(C4) in Lemma 9.3 are all satisfied. If so, we immediately have the theorem from the lemma by setting $\varepsilon = (\log \log n)/n$.

Since Q is determined by components, Q holds for the union S of the solutions S_i of the components of $G - C$. Therefore the condition (C1) is satisfied.

Next consider a maximum set S^* inducing a subgraph with Q in G. Since Q is hereditary, $S' = S^* - C$ induces a subgraph satisfying Q in G, and furthermore each $S_i' = S^* \cap V_i$ induces a subgraph of $G_i = (V_i, E_i)$ satisfying Q. Thus we have $|S_i| \geq |S_i'|$ since S_i is a maximum set of V_i inducing a subgraph with Q. Therefore we have

$$|S| = \sum |S_i| \geq \sum |S_i'| = |S'| \geq |S^*| - |C|,$$

so the condition (C2) is satisfied.

Since at least one nonempty graph satisfies the hereditary property Q, K_1 (the single vertex graph) satisfies Q. Since Q is determined by the components, a maximum independent set I^* of the planar graph G satisfies Q. Therefore

$$|S^*| \geq |I^*| \geq n/4.$$

The last inequality follows from the four color theorem. Thus the condition (C3) is satisfied. The condition (C4) is identical to (c).   Q.E.D.

If Q is recognizable in polynomial-time instead of linear time, there exists a polynomial-time approximation algorithm with the same worst-case ratio although the time complexity is no longer $O(n \log n)$. The following corollary is an immediate consequence of Theorem 9.7.

Corollary 9.1. *For planar graphs, there exist approximation algorithms with time complexity* $O(n \log n)$ *and worst-case ratio* $1 - O(1/(\log \log n)^{1/2})$ *for the maximum induced subgraph problems with respect to the following properties among others*:

(1) *independent*;
(2) *bipartite*;
(3) *forest; and*
(4) *outerplanar*.

9.5. Maximum matching

The *maximum matching problem* asks for a maximum number of pairwise nonadjacent edges in a graph. Although there exists an $O(n^{1.5})$ exact algorithm for the maximum matching problem on planar graphs [LT80, MV80], the efficient approximation algorithm would be useful if the available computation time is limited and an approximate maximum matching is sufficient for some practical purpose. In this section, using algorithm MISP, we give an $O(n \log n)$ approximation algorithm for this problem [CNS81a].

The problem is an instance of the *maximum subgraph problem with property* Q, which is defined similarly as the maximum *induced* subgraph problem in the preceding section. The assertion on the maximum subgraph problem similar to Theorem 9.7 does not always hold true. However one can easily see that the conditions (C1)–(C4) are also sufficient for the success of Algorithm MISP to the maximum matching problem, where S and S^* should be subsets of the edge set E (instead of V) of a given graph $G = (V, E)$. Clearly the conditions (C1) and (C4) are satisfied, and (C2) is also satisfied as will be shown later. However (C3) is not satisfied since a planar graph does not always contain a maximum matching of linear size as $K_{1,n-1}$ or $K_{2,n-2}$ indicate. Thus a direct application of MISP cannot guarantee the worst-case ratio $1 - O(1/(\log \log n)^{1/2})$. However if the problem on a general planar graph can be reduced to the same problem on a particular planar graph having a matching of linear size, for example a planar graph with minimum degree 3 (see Theorem 1.8), then a modified MISP may guarantee the desired worst-case ratio.

The following is the approximation algorithm MATCHING for the maximum matching problem on planar graphs, which is based on the idea above. The algorithm finds a matching $S(G)$ of a given graph G. We denote by $S^*(G)$ a maximum matching of graph G.

```
procedure MATCHING;
  procedure REDUCE(G, S(G));
    begin
      let G contain N vertices, and let v be a vertex of minimum degree of G;
      if N ≤ log log n
        then find a maximum matching S*(G) of G by applying any reason-
             able (polynomial or exponential time) exact algorithm and let
             S(G) := S*(G)
      else if d(v) < 3
        then if d(v) = 0
          then
            begin
              G' := G − v;
              REDUCE(G', S(G'));
              S(G) := S(G')
            end
          else if d(v) = 1
            then
```

```
        begin
          let u be the vertex adjacent to v;
          G' := G − {u, v};
          REDUCE(G', S(G'));
          S(G) := S(G') ∪ {(v, u)}
        end
      else {d(v) = 2}
        begin
          let u, w be the vertices adjacent to v;
          let G' be the graph obtained from G by identifying the
          three vertices v, u and w;
          REDUCE(G', S(G'));
          if S(G') contains no edge which was adjacent to u in G
            then S(G) := S(G') ∪ {(v, u)}
            else S(G) := S(G') ∪ {(v, w)};
        end
    else {d(v) ≥ 3 and N > log log n}
      apply Algorithm MISP with ε = (log log n)/N to G to obtain a match-
      ing S(G)
  end {REDUCE};
begin
  let G be a given planar graph of n vertices;
  REDUCE(G, S(G))
end.
```

Theorem 9.8. [*CNS*81*a*]. *The approximation algorithm MATCHING for the maximum matching problem on planar graphs has a worst-case ratio* $1 - O(1/(\log\log n)^{1/2})$ *and time complexity* $O(n \log n)$.

Proof. (a) Correctness and worst-case ratio

The algorithm MATCHING reduces a graph G to a smaller one G' whenever G contains more than $\log\log n$ vertices including a vertex of degree 2 or less. So one eventually arrives at a graph G_0 which either has N ($\leq \log\log n$) vertices or is of minimum degree ≥ 3.

We first show that MATCHING correctly finds a matching of the graph G_0 within the desired worst-case ratio. If $N \leq \log\log n$, then MATCHING finds a maximum matching of G_0 by applying an exact algorithm, so $|S(G_0)|/|S^*(G_0)| = 1$. (Note that we set $|S(G)|/|S^*(G)|$ to 1 if $|S^*(G)| = 0$.) Otherwise (that is, if $N > \log\log n$ and G_0 is of minimum degree ≥ 3) MATCHING applies Algorithm MISP with $\varepsilon = (\log\log n)/N$ to find a match-ing $S(G_0)$. We shall show that the conditions (C1)–(C4) of Lemma 9.3 are all satisfied in this case. Since G_0 is a planar graph with minimum degree $\geqslant 3$, Theorem 1.8 implies $|S^*(G_0)| \geqslant N/3$, so the condition (C3) is satisfied. Clearly the conditions (C1) and (C4) are satisfied. Finally (C2) is satisfied: all the edges of $S^*(G_0)$ not contained in components of $G_0 - C$ are adjacent to vertices of C; furthermore each vertex of C is adjacent to at most one edge in $S^*(G_0)$; therefore $|S^*(G_0)| - |S(G_0)| \leqslant |C|$. Thus in this case Algorithm MISP with $\varepsilon = (\log \log n)/N$ correctly finds a matching $S(G_0)$ within the worst-case ratio $1 - O(1/(\log \log n)^{1/2})$.

We shall next show that MATCHING correctly forms an approximate matching $S(G)$ of G from a matching $S(G')$ of a smaller graph G' while preserving the desired worst-case ratio, that is, $|S(G)|/|S^*(G)| \geqslant |S(G')|/|S^*(G')|$. Consider the following three cases according to the employed reductions.

Case 1: G contains an isolated vertex v, i.e., a vertex of degree 0.
If $S(G')$ is a matching of $G' = G - v$, then $S(G) = S(G')$ is a matching of G. Furthermore $|S(G)|/|S^*(G)| = |S(G')|/|S^*(G')|$.

Case 2: G contains a vertex v of degree 1.
Let u be the vertex adjacent to v in G, and let $G' = G - \{u, v\}$. If $S(G')$ is a matching of G', then clearly $S(G) = S(G') \cup \{(u, v)\}$ is also a matching of G. Thus $|S(G)| = |S(G')| + 1$. Since a maximum matching $S^*(G)$ of G contains exactly one edge adjacent to u, we have $|S^*(G)| \leqslant |S^*(G')| + 1$. Therefore we have $|S(G)|/|S^*(G)| \geqslant |S(G')|/|S^*(G')|$.

Case 3: G contains a vertex of degree 2.
Let u, w be the vertices adjacent to v, and let G' be the graph obtained from G by identifying u, v, and w into a single vertex. Let $S(G) = S(G') \cup \{(v, u)\}$ if a matching $S(G')$ of G contains no edge which was adjacent to u in G, and otherwise let $S(G) = S(G') \cup \{(v, w)\}$. Clearly $S(G)$ is a matching of G if $S(G')$ is so in G'. Thus we have $|S(G)| = |S(G')| + 1$. If a maximum matching $S^*(G)$ of G contains either (v, u) or (v, w), say (v, u), then $S^*(G) - (v, u)$ is a matching of G'. Otherwise, $S^*(G)$ contains an edge (u, x) with $x \neq v$, and then $S^*(G) - (u, x)$ is a matching of G'. Thus we have $|S^*(G)| \leqslant |S^*(G')| + 1$. These two equations imply $|S(G)|/|S^*(G)| \geqslant |S(G')|/|S^*(G')|$.

Thus we have verified the correctness and the worst-case ratio.

(b) Time complexity

First consider the computation time required for the graph G_0. Let G_0 contain N vertices. If $N \leqslant \log \log n$, MATCHING applies to G_0 an exact algorithm. Even if the exact algorithm has time complexity $O(N2^N)$, it requires at most $O(n \log n)$ time. If $N > \log \log n$ and G_0 is of minimum degree $\geqslant 3$, then the algorithm MISP with $\varepsilon = (\log \log n)/N$ applied to G_0 requires at most $O(\max\{N \log N, N2^{\varepsilon N}\}) = O(N \log n) \leqslant O(n \log n)$ time.

Next consider the time required for reducing a given graph G to the graph G_0. The exhaustive operations are vertex-deletions and vertex-identifications appeared in the reductions. Since one can execute a single vertex-deletion of a vertex v in $O(d(v))$ time using the adjacency lists of a graph, and every vertex appears in at most one vertex-deletion, all the vertex-deletions involved in the execution of MATCHING require at most $O(n)$ time. On the other hand, any sequence of vertex-identifications can be executed in $O(n \log n)$ time by using one of the following devices: adjacency lists together with an adjacency matrix (see Chapter 2); or AVL trees together with an efficient list merging algorithm [Shi80].

Finally it is clear that the time required for forming $S(G)$ from $S(G')$ is, in total, as much as the time required for the reductions. Thus we can implement the algorithm MATCHING to run in at most $O(n \log n)$ time.   Q.E.D.

9.6. Minimum vertex cover

The *minimum vertex cover problem* asks for a minimum number of vertices such that every edge of a graph is incident on at least one of the vertices. The problem is *NP*-complete even for planar graphs. In this section we give an approximation algorithm for this problem.

The minimum vertex cover problem and the maximum independent set problem are quite closely related. For a graph $G = (V, E)$, a vertex set $V' \subset V$ is a (maximum) independent set of G if and only if the complementary set $V - V'$ is a (minimum) vertex cover set of G. However this transformation does not preserve the worst-case ratio [GJ79, p. 134]. So we must design an approximation algorithm for each problem. We will show that the reduction given in Section 9.5 is useful also for the minimum vertex cover problem. The following is the algorithm for the minimum vertex cover problem on planar graphs.

procedure COVER;
  procedure REDUCE(G, $C(G)$);
    begin
      let G be contain N vertices, and let v be a vertex of minimum degree of G;
      if $N \leqslant \log \log n$
        then obtain a minimum vertex cover set $C^*(G)$ of G by checking

every subset of the vertices and let $C(G) := C^*(G)$
else if $d(v) < 3$
  then
    if $d(v) = 0$
      then
        begin
          $G' := G - v;$
          REDUCE($G', C(G')$);
          $C(G) := C(G')$
        end
      else
        if $d(v) = 1$
          then
            begin
              let u be the vertex adjacent to v;
              $G' := G - \{u - v\};$
              REDUCE($G', C(G')$);
              $C(G) := C(G') + u$
            end
          else $\{d(v) = 2\}$
            begin
               let u, w be the vertices adjacent to v;
              if $(u, v) \in E$ *then*
                begin
                  $G' := G - \{u, v, w\};$
                  REDUCE($G', C(G')$);
                  $C(G) := C(G') + \{u, w\}$
                end
              else $\{(u, w) \notin E\}$
                begin
                  let G' be the graph obtained from G by identifying
                  the three vertices v, u and w into a new vertex x;
                  REDUCE($G', C(G')$);
                  if $x \in C(G')$
                    then $C(G) := C(G') - x + \{u, w\}$
                    else $C(G) := C(G') + v$;
                end
             end
  else $\{d(v) \geq 3$ and $N > \log \log n\}$
    begin
      apply the Lipton and Tarjan's algorithm with $\varepsilon = \log \log n/N$ to

find an approximate independent set $I(G)$ of G;
$C(G) := V - I(G)$
end;
end {REDUCE};
begin
let G be a given planar graph with n vertices;
REDUCE$(G, C(G))$
end.

Theorem 9.9. *The approximation algorithm COVER for the minimum vertex cover problem has a worst-case ratio* $1 + O(1/(\log \log n)^{1/2})$ *and time complexity* $O(n \log n)$.

Proof. Clearly the algorithm has time complexity $O(n \log n)$, which is the same as MATCHING. In the same way as for the proof of Theorem 9.8, one can verify the correctness and the worst-case ratio of the algorithm. Consider only the case when G_0 is of $N \geqslant \log \log n$ vertices and with minimum degree 3 or more. In this case by Corollary 1.5 a minimum vertex cover $C^*(G_0)$ of G_0 contains at least $N/3$ vertices. Hence

$$|C(G_0)|/|C^*(G_0)| = (N - |I(G_0)|)/(N - |I^*(G_0)|)$$

$$= 1 + (|I^*(G_0)| - |I(G_0)|)/(N - |I^*(G_0)|)$$

$$\leqslant 1 + O(1/(\log \log n)^{1/2}).$$

The last inequality follows from $|C^*(G_0)| = N - |I^*(G_0)| \geqslant N/3$ and $|I^*(G_0)| - |I(G_0)| \leqslant |C| = O(N/(\log \log n)^{1/2})$, which is ensured by Lipton and Tarjan's algorithm with $\varepsilon = \log \log n/N$. Thus in this case the algorithm COVER correctly finds a vertex cover within the desired worst-case ratio. Furthermore each reduction clearly preserves the worst-case ratio, that is, $|C(G)|/|C^*(G)| \leqslant |C(G')|/|C^*(G')|$.   Q.E.D.

Remark. The planar separator theorem has many applications beyond those presented in this chapter. The layouts of graphs, such as trees, X-trees and k-dimensional meshes, for VLSI are discussed in [Lei80]. Generalization of the "nested dissection" method for carrying out sparse Gaussian elimination on a system of linear equations is discussed in [LRT79]. Applications to the problems of nonserial dynamic programming, pebbling, lower bounds on Boolean circuit size and embedding of data structures are found in [LT80]. Algorithms for the shortest path problems in planar graphs are presented in [Fre83, Fre85, LRT79]. Applications to the planar flow problem are given in [Fre83, Fre85, JV82, HJ85].

CHAPTER 10

HAMILTONIAN CYCLES

10.1. Introduction

A *Hamiltonian cycle* (*path*) of a graph G is a cycle (path) which contains all the vertices of G. The Hamiltonian cycle problem asks whether a given graph contains a Hamiltonian cycle. It is *NP*-complete even for 3-connected planar graphs [Gar76, Kar72]. However the problem becomes polynomial-time solvable for 4-connected planar graphs: Tutte proved that such a graph necessarily contains a Hamiltonian cycle [Tut56, Tut77]. Based on Tutte's proof Gouyou-Beauchamps reported an $O(n^3)$ algorithm which actually finds a Hamiltonian cycle in such a graph [Gou82]. Asano, Kikuchi and Saito presented a linear algorithm for the problem on 4-connected "maximal" planar graphs [AKS84]. Recently Chiba and Nishizeki [CN84] constructed a linear algorithm for 4-connected planar graphs, based on Thomassen's short proof for Tutte's theorem [Tho83, CN85b].

Both Tutte's and Thomassen's proofs use an inductive argument like the following: they first decompose a graph into several small subgraphs, then find Hamiltonian cycles in subgraphs by applying the inductive hypothesis to them, and finally combine the Hamiltonian cycles into a Hamiltonian cycle of the whole graph. The proofs immediately yield a divide-and-conquer algorithm. However, since the decomposed subgraphs are not always (edge-)disjoint, it was nontrivial to verify even the polynomial boundedness of the algorithm.

In this chapter we first present a proof for Tutte's theorem given in [CN84], which is based on Thomassen's proof but avoids the decomposition into nondisjoint subgraphs. The proof is constructive, and yields a simple algorithm for finding a Hamiltonian cycle. We next show that at most $O(n)$ recursive calls occur during one execution of the algorithm, and hence the algorithm runs in $O(n^2)$ time. We then discuss briefly the linear implementation of the algorithm.

Finally in Section 10.4 we discuss a related topic, a "Hamiltonian walk."

10.2. Proof of Tutte's theorem

In this section we define some terms and present a proof for Tutte's theorem.

We first begin with some terms related to a graph decomposition. Let $\{x, y\}$ be a separation pair and let G_1 and G_2 be the split graphs. Remember the definition of a separation pair and split graphs, given in Section 1.2. We sometimes say "split G_1 from G" when constructing G_2 from G (see Fig. 1.6). Note that no multiple edges are produced in G_1 or G_2. The edge (x, y) in G_1 or G_2 is called a *virtual edge* no matter whether it originally exists in G or is newly added to G_1 or G_2. Dividing a graph G into two split graphs G_1 and G_2 is called *splitting*. Reassembling the two split graphs G_1 and G_2 into G is called *merging*. Merging is the inverse of splitting. We define a *separation triple* as a separating set of three vertices. A graph G is *4-connected* if G is 3-connected and has no separation triples.

We are now ready to present Thomassen's result.

Lemma 10.1 (*Tho* 83). *Let G be a 2-connected plane graph with the outer facial cycle Z. Let s and $e = (a, b)$ be a vertex and an edge, both on Z, and let t be any vertex of G distinct from s. Then G has a path P going from s to t through e such that*

(i) *each component of $G - P$ is adjacent to at most three vertices of P, and*

(ii) *each component of $G - P$ is adjacent to at most two vertices of P if it contains a vertex of Z.*

Clearly Lemma 10.1 implies that a 4-connected planar graph G has a Hamiltonian cycle: let s and t be two adjacent vertices on Z and let $e \neq (s, t)$ be an edge on Z, then the path P joining s and t through e, assured by Lemma 10.1, must be a Hamiltonian path of G, so $P + (s, t)$ must be a Hamiltonian cycle of G. Thus an algorithm for finding the s-t path P immediately yields an algorithm for finding a Hamiltonian cycle.

Although the original graph G is 4-connected, the subgraphs into which G is decomposed are no longer 4-connected. However they inherit some favorable property, which we call "internally 4-connected." Intuitively a graph is internally 4-connected if it contains no separation pair or triple in the interior. We formally define the term as follows.

Let G be a 2-connected plane graph with outer facial cycle Z. Let s and t be two distinct vertices on Z, and let $e = (a, b)$ be an arbitrary edge on Z such that $e \neq (s, t)$. Interchanging the roles of s and t or a and b and also mirroring the plane embedding of G if necessary, one may assume without loss of generality that $t \neq a, b$ and vertices s, a, b and t appear clockwise on Z in this order. (See Fig. 10.1(a).) Note that possibly $s = a$ as illustrated in Fig. 10.1(b). Let r be the vertex on Z counterclockwise next to s, and let $f = (r, s)$ be the edge joining r and s. If vertices x and y are on Z, P_{xy} denotes the "outer" path going from x to y clockwise on Z. Then a plane 2-connected graph G is *internally 4-connected with respect to* (s, t, e) if G satisfies the following two conditions.

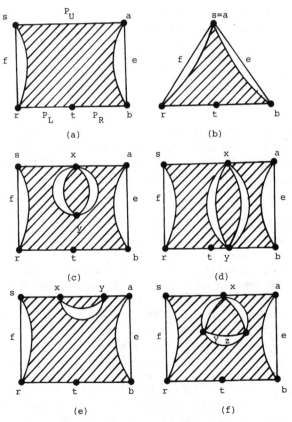

Fig. 10.1. Illustrations for notations and conditions (a) and (b): (a) shows G with $s \neq a$; (b) shows G with $s = a$; (c), (d) and (e) show separation pairs violating conditions (a); and (f) shows a separation triple $\{x, y, z\}$ violating condition (b).

(a) If $\{x, y\}$ is a separation pair, then (i) every component of $G - \{x, y\}$ contains at least one vertex of Z and (ii) each of the three paths P_{sa}, P_{bt} and P_{ts} contains at most one of x and y.

(b) If $\{x, y, z\}$ is a separation triple, then every component of $G - \{x, y, z\}$ contains at least one vertex of Z.

Condition (a)(i) above implies that both x and y of the separation pair must lie on Z and that $G - \{x, y\}$ contains exactly two components, while condition (b) implies that at least two of x, y and z of the separation triple must lie on Z.

Figs. 10.1(c), (d) and (e) show separation pairs violating condition (a), while Fig. 10.1(f) shows a separation triple violating condition (b). An internally 4-connected graph is diagrammatically illustrated in Fig. 10.2(a). A separation pair $\{x, y\}$ is called *vertical* if either $x \in P_{sa} - s$ and $y \in P_{br}$ or $x = s$ and $y \in P_{bt} - t$. In Fig. 10.2(a) $\{x, y\}$ is one of the vertical separation pairs. In the algorithm, G is decomposed into two subgraphs with respect to a vertical separation pair. (Such a decomposition is called a Type I reduction.) If $\{x, y\}$ is not vertical, one of x and y must be in $P_{bt} - t$ and the other in $P_{tr} - t$, and such a pair is implicitly called *horizontal*. In Fig. 10.2(a) $\{y', z'\}$ is a horizontal separation pair.

Let G be an internally 4-connected plane graph having no vertical separation pair. (Fig. 10.3(a) illustrates such a graph G with $s \neq a$, while Fig. 10.3(b) illustrates G with $s = a$.) Then G is decomposed into several subgraphs, called G_b, G_g^2, G_u^3 and G_g^4, which are defined below. (If $s = a$, G is decomposed into G_b and G_g^2.)

Let C_b be the block of $G - P_{sa}$ which contains t. (In Figs. 10.3(a) and (b) C_b are cross-hatched.) Then C_b must entirely contain P_{br}, otherwise G would contain a vertical separation pair. Repeat splitting of C_b at every separation pair such that one of the two split graphs entirely contains P_{br}. (In case of Fig. 10.3(a) C_b is split into four components, as illustrated in Fig. 10.3(c).) The resulting components are called G_b or G_g^2. That is, the component containing P_{br} is called G_b, while each of the others is called G_g^2 where $g = (x, y)$ is a virtual edge contained in the component. Here we may assume by possibly interchanging the roles of x and y that $x \neq b$.

We next define G_u^3 for each cutvertex u of $G - P_{sa}$ contained in C_b. (There are six u's in Fig. 10.3(a).) Let C_u be the maximal subgraph of $G - P_{sa}$ which can be separated from C_b at u. Let C_u^3 be the subgraph of G induced by the vertices of C_u and the vertices on P_{sa} which are adjacent to C_u. Let x (resp. y) be the vertex of $P_{sa} \cap C_u^3$ which is nearest to s (resp. a) along P_{sa}. Add a new edge $e' = (u, y)$ to C_u^3 if it does not exist, and let G_u^3 be the resulting graph. Fig. 10.3(d) illustrates G_u^3.

Finally we define G_g^4 for a virtual edge g. Schematically G_g^4 is a graph formed

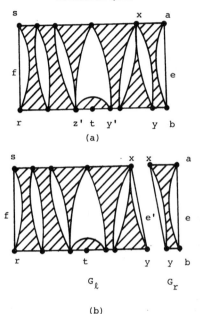

(a)

(b)

Fig. 10.2. Type I reduction for an internally 4-connected graph G having vertical separation pairs; (a) G; (b) G_l and G_r.

by merging G_g^2 with all the $G_{g'}^2$ and C_u^3 drawn above G_g^2. Formally we constructively define G_g^4 as follows (see Fig. 10.3(e)):

(1) $C_g^4 := G_g^2$;
(2) while C_g^4 has a virtual edge $g' \neq g$, iterate to merge $G_{g'}^2$ into C_g^4;
(3) for each cutvertex $u(\neq x, y)$ of $G - P_{sa}$ contained in C_g^4, merge C_u^3 to C_g^4;
(4) construct the subgraph of G induced by the vertices of C_g^4 together with the vertices of P_{sa} adjacent to C_g^4, and redefine C_g^4 as the subgraph;
(5) $G_g^4 := C_g^4 - x - y$.

Fig. 10.3(e) illustrates $G_{g_2}^4$ for virtual edge g_2. Let v (resp. w) be the vertex of G_g^4 that appears first (resp. last) on P_{sa}. Let w' (resp. v') be the vertex which is adjacent to y (resp. x) and appear last (resp. first) on the outer path P_{wv} of G_g^4.

We now have the following lemma.

Lemma 10.2. *Let G be an internally 4-connected plane graph containing no vertical separation pair. Then all the decomposed graphs G_b, G_g^2, G_u^3 and G_g^4 are internally 4-connected with respect to (s', t', e') if s', t' and e' are defined as follows.*

(a) *Case G_b: Let $s' = b$, and $t' = t$. If $t \neq r$, let e' be the edge clockwise incident*

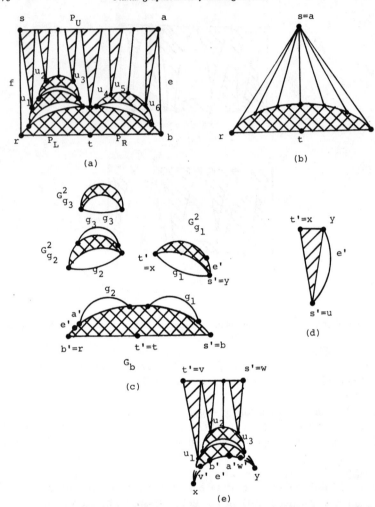

Fig. 10.3. Type II reduction for an internally 4-connected graph G having no vertical
separation pair: (a) G with $s \neq a$; (b) G with $s = a$; (c) G_b and G_g^2 where
g_1, g_2 and g_3 are virtual edges; (d) G_u^3; (e) G_g^4.

to r on the outer facial cycle Z' of G_b; if $t = r$, let e' be the edge
counterclockwise incident to b.

(b) *Case G_g^2: Let $s' = y$, $t' = x$, and let e' be the edge counterclockwise incident
to y on the outer facial cycle of G_g^2.*

(c) *Case G_u^3*: Let $s' = u$, $t' = x$, and $e' = (u, y)$.

(d) *Case G_g^4*: Let $s' = w$, and $t' = v$. If $w' \neq v'$, let $e' = (a', b')$ be an arbitrary edge on outer path $P_{w'v'}$; if $w' = v'$, let e' be an arbitrary edge on P_{wv} incident to $w'(= v')$.

Proof. Since the proof for cases (a)–(c) is trivial, we verify only the case (d). Suppose that G_g^4 is not internally 4-connected with respect to (s', t', e'). Then G_g^4 must have one of the following three:

(i) a cut vertex v_c;

(ii) a separation pair $\{v_{p1}, v_{p2}\}$ such that both v_{p1} and v_{p2} are contained in path $P_{wa'}$ or $P_{b'v}$; and

(iii) a separation triple $\{v_{t1}, v_{t2}, v_{t3}\}$ for which one of the components of $G_g^4 - \{v_{t1}, v_{t2}, v_{t3}\}$ contains no vertex of the outer facial cycle Z' of G_g^4.

In either case G would contain a separation triple $\{v_c, x, y\}$, $\{v_{p1}, v_{p2}, x\}$, $\{v_{p1}, v_{p2}, y\}$, or $\{v_{t1}, v_{t2}, v_{t3}\}$ such that the deletion of the triple from G produces a component containing no vertex of Z. This contradicts the internally 4-connectedness of G. Q.E.D.

Lemma 10.3 below claims that a plane graph internally 4-connected with respect to (s, t, e) has a Hamiltonian path joining s and t. Although the claim is implied by Lemma 10.1, we give a constructive proof, from which a simple algorithm for finding a Hamiltonian path follows immediately.

Lemma 10.3 (*CN84*). *Let G be a plane graph having an outer facial cycle Z. Let s and t be two distinct vertices on Z and let $e(\neq (s, t))$ be an edge on Z. If G is internally 4-connected with respect to (s, t, e), then G has a Hamiltonian path $P(G, s, t, e)$ which connects s and t and contains e. Moreover, if G has no vertical separation pair, then $P(G, s, t, e)$ does not contain edge $f = (s, r)$.*

Proof. The proof is by induction on the number n of vertices of a graph G. If $n = 3$, the claim is clearly true, so we assume that $n > 3$. There are two cases to consider.

Case 1: there exists a vertical separation pair $\{x, y\}$. (See Fig. 10.2.) Denote by G_l and G_r the two $\{x, y\}$-split graphs. One may assume that G_r contains e. Among all vertical separation pairs of G, we choose $\{x, y\}$ such that G_r has the smallest number of vertices. We call such a separation pair the *rightmost separation pair*.

First consider the case $t \in G_l$ as shown in Fig. 10.2. Let $e' = (x, y)$ be a virtual edge. Then clearly G_l is internally 4-connected with respect to (s, t, e'), while G_r is internally 4-connected with respect to (x, y, e) (or with respect to

(y, x, e) if $y = b$). Therefore by the inductive hypothesis G_l has a Hamiltonian path $P(G_l, s, t, e')$ and G_r has $P(G_r, x, y, e)$ (or $P(G_r, y, x, e)$). Thus G has a Hamiltonian path

$$P(G, s, t, e) = P(G_l, s, t, e') + P(G_r, x, y, e) - e'$$

or

$$P(G_l, s, t, e') + P(G_r, y, x, e) - e'.$$

Next consider the case $t \notin G_l$. Let $e' = (y, x)$, then G_l and G_r are internally 4-connected with respect to (y, s, e') and (x, t, e), respectively. Here $s \neq x$ since G is internally 4-connected with respect to (s, t, e). Thus G has a Hamiltonian path

$$P(G, s, t, e) = P(G_l, y, s, e') + P(G_r, x, t, e) - e'$$

as desired. Note that $P(G_r, x, t, e)$ does not contain (x, y) since G_r has no vertical separation pair.

Case 2: G has no vertical separation pairs (see Fig. 10.3). By the inductive hypothesis and Lemma 10.2, all the decomposed graphs G_b, G_g^2, G_u^3 and G_g^4 have Hamiltonian paths. From them one can construct a Hamiltonian path $P(G, s, t, e)$ of G as follows.

(i) First set P to $P_{sb} + P(G_b, b, t, e')$.
(ii) Then P is extended into a Hamiltonian path $P(G, s, t, e)$ of whole G. Note that in the following extensions edge $f = (s, r)$ is never included in path P.
    (ii-1) While there is a virtual edge $g = (x, y)$ in G_b, repeat the following extension of P:
        if $g \in P$, then let $P := P - g + P(G_g^2, y, x, e')$ (that is, replace g of P by a Hamiltonian path of G_g^2), and subsequently merge G_g^2 into G_b;
        if $g \notin P$, then construct G_g^4, and set $P := P - P_{vw} + P(G_g^4, w, v, e')$, and subsequently delete g from G_b.
    (ii-2) Finally, for each u of the cutvertices of $G - P_{sa}$ contained in G_b, set $P := P - P_{xy} + P(G_u^3, u, x, (u, y)) - (u, y)$.

Clearly the resulting P forms a Hamiltonian path $P(G, s, t, e)$ of the whole G.
                                                                            Q.E.D.

Example 10.1. For an illustration consider G in Fig. 10.3(a) which has no vertical separation pair. If $P(G_b, b, t, e')$ contains virtual edge g_1 but not g_2, then $P(G, s, t, e)$ is constructed from Hamiltonian paths of G_b, $G_{g_1}^2$, $G_{u_4}^3$, $G_{u_5}^3$, $G_{u_6}^3$ and $G_{g_2}^4$. If $P(G_b, b, t, e')$ contains both g_1 and g_2 and if $P(G_{g_2}^2, s', t', e')$ does not contain g_3, then $P(G, s, t, e)$ is constructed from Hamiltonian paths of G_b, $G_{g_1}^2$, $G_{g_2}^2$, $G_{g_3}^4$, $G_{u_1}^3$, $G_{u_4}^3$, $G_{u_5}^3$ and $G_{u_6}^3$.

We next give another concrete example.

Example 10.2. Consider an internally 4-connected planar graph G in Fig. 10.4(a) which has no vertical separation pair. The block C_b is drawn in a dotted circle. All the decomposed graphs G_b, $G_{(1,3)}^2$, $G_{(1,7)}^2$, $G_{(3,5)}^2$, $G_{(5,7)}^2$, G_2^3, G_4^3 and G_6^3 are depicted in Fig. 10.4(b) where virtual edges are drawn in dotted lines.

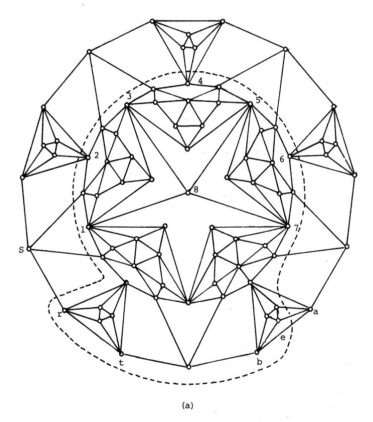

(a)

Fig. 10.4. (a) An internally 4-connected planar graph G; (b) decomposed graphs; (c) $G_{(3,5)}^4$ and (d) a Hamiltonian path of G.

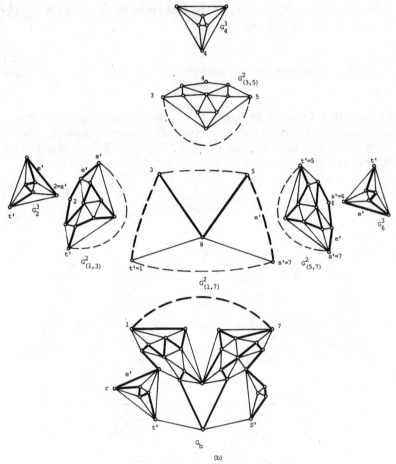

(b)

Fig. 10.4 (continued)

Suppose that the Hamiltonian path of G_b indicated by thick lines is determined by the inductive hypothesis. Since the path passes through the virtual edge $(1, 7)$, a Hamiltonian path of $G_{(1,7)}^2$ is determined. Suppose that the path passes through virtual edges $(1, 3)$ and $(5, 7)$ but not $(3, 5)$ as indicated by thick lines. Then Hamiltonian paths of $G_{(1,3)}^2$, G_2^3, $G_{(5,7)}^2$, G_6^3, and $G_{(3,5)}^4$ are determined. The first four paths are indicated by thick lines in Fig. 10.4(b), while the last path in $G_{(3,5)}^4$ is depicted in Fig. 10.4(c). Finally combining all these paths, we can get a Hamiltonian path of G as depicted in Fig. 10.4(d).

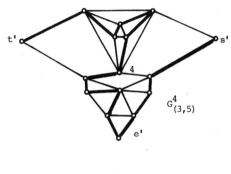

(c)

(d)

Fig. 10.4 (continued)

10.3. Algorithm and $O(n^2)$ **bound**

The proof of Lemma 10.3 immediately yields a recursive algorithm which finds a Hamiltonian path $P(G, s, t, e)$ in an (internally) 4-connected planar graph G. The reduction performed with respect to a vertical separation pair at Case 1 is called a "Type I reduction" (Fig. 10.2), while the reduction at Case 2 is called a "Type II reduction" (Fig. 10.3). Then the algorithm is as follows.

procedure HPATH(G, s, t, e);
  begin
    if G contains exactly three vertices
      then {G is a triangle}
        return a (trivial) Hamiltonian path $P(G, s, t, e)$;
      else if G has a vertical separation pair
        then Type I reduction
        else Type II reduction
  end;

As shown below it is rather easy to verify an $O(n^2)$ time bound for the algorithm HPATH. Clearly the running time of the algorithm is dominated by the time required by the graph decomposition: the decomposition of G into G_l and G_r in Type I reductions; and into G_b, G_g^2, G_u^3 and G_g^4 in Type II reductions. Hopcroft and Tarjan have reported a linear algorithm which decomposes a given graph into 3-connected components [HT73]. Using a similar algorithm, the above decomposition can be done in linear time per reduction. Furthermore we have the following lemma.

Lemma 10.4. *There are at most* $n - 3$ *reductions during one execution of algorithm HPATH.*

Proof. Assume that r reductions occurred during one execution of HPATH. Then we should verify $r \leqslant n - 3$. Let $d(i)$, $1 \leqslant i \leqslant r$, be the number of small graphs into which a graph is decomposed at the ith reduction. That is, $d(i)$ is the number of the recursive calls which occurred at the reduction. Then $d(i)$ is necessarily 2 if the ith reduction is Type I, while $d(i)$ is 1 if the ith reduction is Type II and none of G_g^2, G_g^4 and G_u^3 is produced. Let r' be the number of reductions with $d = 1$. Recall that G is eventually decomposed into triangles, for which Hamiltonian paths can be found trivially and no more reductions occur. Let t be the number of these final triangles.

Consider the so-called recursive call tree. Each internal node of the tree corresponds to a reduction, its sons to the recursive calls at the reduction, and the leaves to the final triangles. Thus the tree has t leaves and r internal nodes, r' of which have outdegree 1. The trivial fact that the number of internal nodes in a binary tree is one less than the number of leaves implies that the recursive call tree has at most $t - 1$ internal nodes of outdegree 2 or more. Thus we have

$$r \leqslant t - 1 + r'. \tag{10.1}$$

A Type I reduction decomposes a graph G into G_l and G_r having two duplicated vertices. The total length (i.e. number of edges) n of Hamiltonian paths in G_l and G_r is one larger than the length $n - 1$ of a Hamiltonian path in G. Thus a Type I reduction increases by one the total length of paths which will be found in the two reduced graphs. In general, the ith reduction increases by at most $d(i) - 1$ the total length of paths which will be found in the $d(i)$ reduced graphs. Conversely, if $d(i) = 1$, then the length of a path which will be found in the reduced graph decreases by at least one. Trivially HPATH initially wishes to find a Hamiltonian path of length $n - 1$. Therefore the total length of paths found in triangles cannot exceed

$$n - 1 + \sum_{1 \leqslant i \leqslant r} (d(i) - 1) - r' = n - 1 + r + t - 1 - r - r'$$

$$= n + t - r' - 2.$$

Since the lengths of the Hamiltonian paths found in the final triangles total $2t$, we have

$$2t \leqslant n + t - r' - 2,$$

that is,

$$t \leqslant n - r' - 2. \tag{10.2}$$

Combining (10.1) with (10.2), we get the claimed bound on r:

$$r \leqslant n - 3.$$

Q.E.D.

Thus we have the following results.

Theorem 10.1. *The algorithm HPATH spends at most* $O(n^2)$ *time.*

Linear implementation of HPATH is given in [CN84]. The key idea is to use, in place of the Hopcroft and Tarjan's algorithm [HT73], a new algorithm to decompose a plane graph into small subgraphs by traversing some facial cycles. A sophisticated analysis shows that each of the edges is traversed at most constant times during one execution of HPATH, and hence HPATH runs in linear time.

10.4. Hamiltonian walk

A *Hamiltonian walk* of a connected graph is a shortest closed walk that passes through every vertex at least once, and the length of a Hamiltonian walk is the total number of edges traversed by the walk. Obviously a Hamiltonian cycle is a Hamiltonian walk. A trivial lower bound and a trivial upper one are known on the length, $h(G)$, of a Hamiltonian walk of a connected graph G with n vertices: $n \leqslant h(G) \leqslant 2(n-1)$ [GH74]. These bounds follow immediately from the facts: (i) any Hamiltonian walk must pass through every vertex at least once, and (ii) a closed spanning walk traversing twice every edge of any tree of a connected graph is of length $2(n-1)$. However, no nontrivial upper bounds are known even for any interesting class of restricted graphs except a certain class of dense graphs [Ber75, Jol75].

In this section we present without proof a result given by Asano, Nishizeki and Watanabe [ANW80]. They established a nontrivial upper bound on the length of a Hamiltonian walk for maximal planar graphs.

Theorem 10.2. *The length $h(G)$ of a Hamiltonian walk of any maximal planar graph with n vertices satisfies*

$$h(G) \begin{cases} \leqslant 3(n-3)/2 & \text{if } n \geqslant 11, \\ = n & \text{otherwise.} \end{cases}$$

Since the proof in [ANW80] is constructive, it immediately yields an $O(n^2)$ algorithm for actually finding a closed spanning walk of length $\leqslant 3(n-3)/2$ for a given maximal planar graph [NAW83]. The algorithm uses a divide-and-conquer approach involving a partition of a graph at a separation triple. A separation triple forms a triangle in a maximal planar graph. One can improve the time complexity of the algorithm to $O(n)$ by using two algorithms: the linear implementation of HPATH in Section 10.3 and the triangle listing algorithm in Section 8.3. Since $h(G) \geqslant n$, the algorithm has the worst-case ratio $\frac{3}{2}$ as an approximation algorithm for the Hamiltonian walk problem.

CHAPTER 11

FLOWS IN PLANAR GRAPHS

11.1. Introduction

The network flow problem and its variants have been extensively studied. The original and most basic problem is that of finding the maximum flow of a single commodity in an arbitrary directed graph. The key theorem of flow theory is the Max Flow–Min Cut theorem of Ford and Fulkerson [FF56] which holds for single commodity and two-commodity flows [Hu69]. There are efficient algorithms for finding a maximum single-commodity flow; the $O(nm \log n)$ time algorithm of Sleator and Tarjan [Sle80, ST83] is the theoretically best known one for sparse graphs. Two-commodity flows in undirected graphs can be found by solving two single-commodity flow problems, so it can be done in $O(nm \log n)$ time [Ita78, Sak73, Sey79].

The situation is different with regard to flows of more than two commodities. No true polynomial time algorithm is known for the multicommodity flow problem on general graphs. In the multicommodity flow problem, one would like to (1) test the feasibility, that is, decide whether a given graph G has multicommodity flows, each from a source to a sink and of a specified demand, and (2) then actually find them if G has. Like all network flow problems, the multicommodity flow problem can be formulated as a linear program. Thus it can be solved by the simplex or new LP methods [Chv83, Kar84, Kha79]. However, experience has shown that for many specific problems special purpose algorithms can be devised which work better than these methods. A case of the multicommodity flow problem in which the network is planar has a number of important applications, such as traffic control, design of communication networks and routing of VLSI circuits [NSS85]. Several papers have been published on this problem [Dd72, Sak66, Tan64].

We concentrate on planar undirected graphs. It has been established that the Max Flow–Min Cut theorem of multicommodity type holds for the following four classes of planar undirected graphs [OS81, Oka83, Sey81]:

C_1 : all sources and sinks are located on a specified face boundary;

C_{12}: all sources and sinks are located on two specified face boundaries with each source–sink pair on the same boundary;

C_{01}: some source–sink pairs are located on a specified face boundary, and all the other pairs share a common sink located on the boundary (their sources may be located anywhere);

C_a : all the sources can be joined with the corresponding sinks without violating planarity.

Fig. 11.1 illustrates the four classes. Efficient algorithms for the four classes have been reported [MNS85, MNS86, SNS85, Has84]. All the algorithms reduce the flow problem on a planar undirected graph to the shortest path or cycle problem on an undirected or directed graph obtained from the dual of the given undirected graph. In this chapter we present some of them; the so-called uppermost path algorithm for the maximum single-commodity for C_a in Section 11.3, and multicommodity flow algorithms for C_1 in Section 11.4 and for C_a in Section 11.5.

11.2. Definition of multicommodity flows

A *(flow) network* $N = (G, P, c)$ is a triplet, where:

(i) $G = (V, E)$ is a finite undirected graph with vertex set V and edge set E;

(ii) P is the set of *source–sink pairs* (s_i, t_i), where source s_i and sink t_i are distinguished vertices in G.

(iii) $c: E \to R^+$ is the *capacity* function. (R (or R^+) denotes the set of (positive) real numbers.)

Sources and sinks are called *terminals* if distinction is not necesary. A network $N = (G, P, c)$ is *planar* if G is planar, and is a *k-network* if N has k source–sink pairs, that is, $|P| = k$. The four networks in Fig. 11.1 are all 4-networks.

Each source–sink pair (s_i, t_i) of N is given a nonnegative *demand* $d_i \geqslant 0$. Although G is undirected, we orient the edges of G arbitrarily so that the sign of a value of a flow function can indicate the real direction of the flow in an edge. A set of functions $\{ f_1, f_2, \ldots, f_k \}$ with each $f_i : E \to R$ is *k-commodity flows* of demands $d_1, d_2, \ldots, d_k$ if it satisfies:

(a) the *capacity rule*: for each $e \in E$

$$\sum_{i=1}^{k} |f_i(e)| \leqslant c(e);$$

(b) the *conservation rule*: each f_i satisfies

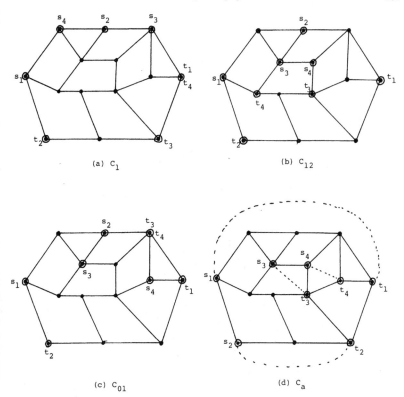

Fig. 11.1. Four classes of planar networks.

$$\text{IN}(f_i, v) = \text{OUT}(f_i, v) \quad \text{for each } v \in V - \{s_i, t_i\},$$

and

$$\text{OUT}(f_i, s_i) - \text{IN}(f_i, s_i) = \text{IN}(f_i, t_i) - \text{OUT}(f_i, t_i) = d_i,$$

where $\text{IN}(f_i, v)$ is the total amount of the flow f_i of commodity i entering v, that is, $\text{IN}(f_i, v) = \Sigma f_i(e) - \Sigma f_i(e')$, the first sum being over all the edges e entering v and with $f_i(e) > 0$, and the second over all the edges e' emanating from v and with $f_i(e') < 0$; $\text{OUT}(f_i, v)$, similarly defined, is the total amount of the flow of commodity i emanating from v.

Consider the 3-network depicted in Fig. 11.2(a) with demands $d_1 = 3$, $d_2 = 4$ and $d_3 = 2$. Three-commodity flows satisfying the demands are illustrated in Fig. 11.3, where arrows indicate actual directions of flows through edges and numbers next to arrows are amounts of flows.

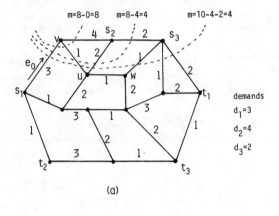

(a)

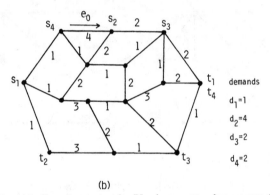

(b)

Fig. 11.2. Two planar networks. (Numbers next to edges are capacities.)

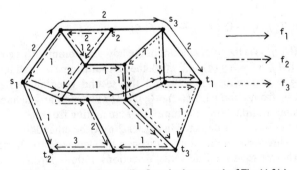

Fig. 11.3. Three-commodity flows in the network of Fig. 11.2(a).

11.3. Planar single-commodity flow

In this section we discuss the single commodity flow problem, that is, the case $k = 1$. The so-called *uppermost path algorithm* can find a maximum single commodity flow in a planar network with source s and sink t both on the outer boundary B, that is, a 1-network in class C_a [FF56, IS79]. The algorithm starts with zero flow and pushes flow as much as possible through the "uppermost path" on B connecting s and t. (The paths are indicated by dotted lines in Fig. 11.4.) Thereby, at least one edge becomes saturated. Such an edge is deleted, and the process is repeated using the uppermost path of the resultant graph.

Uppermost path algorithm
1. Initialize;
   start with zero flow:
   for all $e \in E$ set $f(e) = 0$.
2. Find the uppermost path Q, if none exists then stop.
3. Let $D = \min\{c(e) : e \in Q\}$.
4. Increase the flow by D units along Q:
   $f(e) := f(e) + D$ if $e \in Q$ and the orientation of e is the same as the direction of Q,
   $f(e) := f(e) - D$ if $e \in Q$ and the orientation of e is opposite to the direction of Q.
5. Decrease the capacities on Q:
   $c(e) := c(e) - D$ if $e \in Q$.
6. Delete the edges on Q of zero capacity.
7. Go to 2.

The algorithm is illustrated in Fig. 11.4. In Figs. 11.4(a)–(e) numbers next to edges are their capacities, while in Fig. 11.4(f) numbers are flow values and arrows indicate the real directions of the found flow through edges.

Theorem 11.1. *The uppermost path algorithm finds a maximum flow, whose value is equal to the minimum cut separating s and t.*

Proof. By induction on the number of edges:
   (1) The claim is obvious if the network contains only one edge.
   (2) Assume that the claim is true for all networks having edges less than m, and let N be any network with m edges. Let N' be the resultant network after pushing D units of flow along Q. Since N' contains less than m edges, the algorithm finds flow of value equal to the minimum cut of N'. The minimum

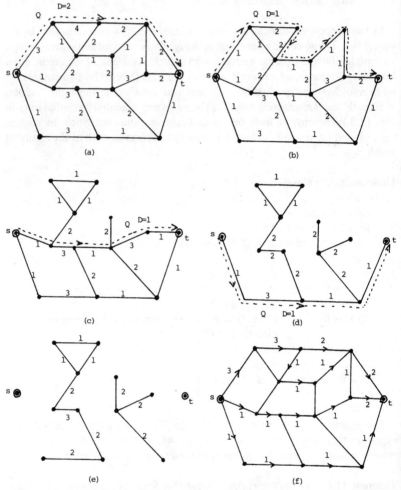

Fig. 11.4. Uppermost path algorithm.

cut of N is larger than that of N' by exactly D. Thus the found flow has value
equal to the minimum cut of N, so is maximum.   Q.E.D.

One can observe that the algorithm merely executes the shortest path
computation on the dual of $G = (V, E)$ [Has81]. Let G_a be the graph obtained
from G by adding an edge joining s and t of infinite capacity, and let

$G_a^* = (V^*, E^*)$ be the dual of G_a. G_a are drawn by solid lines and G_a^* by dotted lines in Fig. 11.5(a). Each oriented edge $e = (i, j)$ in G is associated with an oriented edge $e^* = (i^*, j^*)$ in G_a^*, where the arrowhead of e first touches the arrowhead of e^* when e is rotated clockwise in the plane, as illustrated in Fig. 11.6. It should be noted that both G_a and G_a^* are undirected graphs although the correspondence between their edges are defined as oriented edges. Define a length function l on E^*, such that $l(e^*)$ is equal to the capacity $c(e)$ of the corresponding edge in E. Let $u(v^*)$, $v^* \in V^*$, be the "potential" of v^*, that is, the length of a shortest path from s^* to v^*. The value of $u(v^*)$ is written in the circle next to v^* in Figs. 11.5(a) and (b). A maximum flow function f can simply be constructed as follows: $f(i, j) = u(j^*) - u(i^*)$. The function f is written in Fig. 11.5(b), where a saturated minimum cut is indicated by a dotted line. The proof follows from the following observations:

(1) For every $(i^*, j^*) \in E^*$ on a shortest $s^* t^*$-path, $u(j^*) - u(i^*) = c(i, j)$. Therefore f saturates the minimum cut of N.

(2) For every $(i^*, j^*) \in E^*$, $u(j^*) - u(i^*) \leqslant c(i, j)$. Therefore $f(i, j) \leqslant c(i, j)$ for every $(i, j) \in E$, that is, f satisfies the capacity rule.

(3) For every clockwise cycle $Z \subset E^*$, $\Sigma_{(i^*, j^*) \in Z} (u(j^*) - u(i^*)) = 0$. Therefore for every $v \in V$, $\Sigma_{w \in N(v)} f(v, w) = 0$, that is, f satisfies the conservation rule.

Thus the maximum single-commodity flow can be found in $O(T(n))$ time. Throughout this chapter $T(n)$ denotes the time required for finding the single-source shortest paths in a planar undirected graph with nonnegative edge weights having n vertices, while $T_{all}(n)$ denotes the time for finding all pairs of shortest paths. If the usual Dijkstra's algorithm [AHU74, Joh77] is used, then $T(n) = O(n \log n)$ and $T_{all}(n) = O(n^2 \log n)$. If Frederickson's algorithm [Fre83, Fre85] is used, then either $T(n) = O(n \log^{1/2} n)$ or $T(n) = O(n)$ assuming that the preprocessing is done in $O(n \log n)$ time, and $T_{all}(n) = O(n^2)$.

The uppermost path algorithm above does not work when s and t are not on the same face boundary. Reif [Rei83] and Hassin and Johnson [HJ85] have given $O(T(n) \log n)$ algorithms for such a general undirected planar graph, while Johnson and Venkatesan [JV82] have reported an $O(n^{3/2} \log n)$ algorithm for a general directed planar graph.

11.4. Multicommodity flows for C_1

In this section we discuss the multicommodity flow problem for the class C_1 of undirected planar graphs in which all the sinks and sources are located on one specified face boundary, say the outer face boundary. In this case the uppermost path algorithm in the preceding section cannot be adapted directly to the problem. We present an efficient algorithm MULTIFLOW1 given by Matsumoto, Nishizeki and Saito [MNS85].

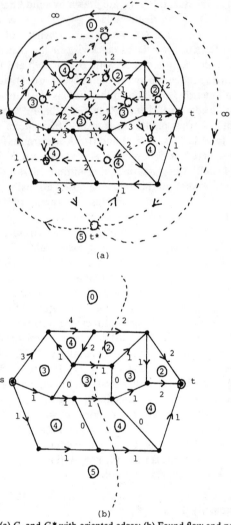

Fig. 11.5. (a) G_a and G_a^* with oriented edges: (b) Found flow and potential.

11.4.1. Preliminary

Although a given planar graph G may be altered during the execution of our algorithm, we generically denote by B the boundary of the outer face of G. We

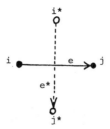

Fig. 11.6. Correspondence of e and e^*.

also use B for the set of edges on B when there is no possibility of confusion. The boundary B is a (simple) cycle if G is 2-connected, but is a closed walk in general. A bridge, i.e. an edge whose deletion disconnects a connected graph, may appear twice in B. We also assume that the vertices on B are $v_0, v_1, \ldots, v_b$, taken in clockwise order. Further we may assume without loss of generality that the given planar graph G is connected.

We denote by $E(X, Y)$ the set of edges with one end in $X \subset V$ and the other in $Y \subset V$. If $X \subset V$, then $E(X) = E(X, V - X)$ is called a *cut*. Especially $E(X)$ is called a *cutset* of G if $G - E(X)$ has exactly two connected components, that is, both $G - X$ and $G - (V - X)$ are connected. Define:

$$c(X, Y) = \sum_{e \in E(X,Y)} c(e);$$

and

$$c(X) = c(X, V - X) \text{ (the capacity of a cut)}.$$

We denote by $d(X, Y)$ the sum of the demands of all source–sink pairs with a source or sink in X and the other in Y. Define $d(X) = d(X, V - X)$. Clearly $E(X) = E(V - X)$, $c(X) = c(V - X)$, and $d(X) = d(V - X)$.

We say that a network N satisfies the *cut condition* for the given demands if $c(X) \geqslant d(X)$ for every $X \subset V$. The cut condition is necessary for the existence of k-commodity flows satisfying given demands in a k-network, but not always sufficient. However Okamura and Seymour have proved the following theorem.

Theorem 11.2. (*OS*81). *Let $N = (G, P, c)$ be a planar k-network, and let all the sources and sinks be on the boundary B of the outer face of a planar graph G, that is, N belongs to class C_1. Then N has k-commodity flows satisfying the demands if and only if N satisfies the cut condition. (A proof will appear later in subsection* 11.4.3.)

It is rather easy to prove the following two lemmas, which hold even if G is not planar.

Lemma 11.1. *Let $G = (V, E)$ be a connected graph. A network $N = (G, P, c)$ satisfies the cut condition if and only if $c(X) \geqslant d(X)$ for each cutset $E(X)$.*

Proof. Since the necessity is obvious, we shall prove the sufficiency. Suppose that $c(X) \geqslant d(X)$ for every cutset $E(X)$, and that there is a subset X of V such that $c(X) < d(X)$. Then either $G - X$ or $G - (V - X)$ is disconnected since $E(X)$ must not be a cutset. Choose such a subset X that $|E(X)|$ is minimum. Clearly $|E(X)| \geqslant 1$. We may assume without loss of generality that $G - X$ is disconnected. Then there exist $Y, Z \subset V$ such that $V - X = Y \cup Z$, $Y \cap Z = \varnothing$ and $E(Y) \cap E(Z) = \varnothing$. Obviously $c(X) = c(Y) + c(Z)$. Since $|E(Y)|, |E(Z)| < |E(X)|$, by the choice of $E(X)$ we have $c(Y) \geqslant d(Y)$, $c(Z) \geqslant d(Z)$. Moreover $d(X) \leqslant d(Y) + d(Z)$. Thus we have $c(X) \geqslant d(X)$, a contradiction.   Q.E.D.

Lemma 11.2. *If $G = (V, E)$ is a graph and $X, Y \subset V$, then*

$$c(X \cap Y) + c(X \cup Y) = c(X) + c(Y) - 2c(X - Y, Y - X);$$

and

$$d(X \cap Y) + d(X \cup Y) = d(X) + d(Y) - 2d(X - Y, Y - X).$$

Proof. Clearly

$$c(X \cap Y) = c(X \cap Y, X - Y) + c(X \cap Y, Y - X)$$
$$+ c(X \cap Y, V - X \cup Y),$$

and

$$c(X \cup Y) = c(X - Y, V - X \cup Y) + c(Y - X, V - X \cup Y)$$
$$+ c(X \cap Y, V - X \cup Y).$$

Therefore we have

$$c(X \cap Y) + c(X \cup Y)$$
$$= \{c(X - Y, V - X \cup Y) + c(X \cap Y, V - X \cup Y)$$
$$+ c(X \cap Y, Y - X) + c(X - Y, Y - X)\}$$
$$+ \{c(Y - X, V - X \cup Y) + c(X \cap Y, V - X \cup Y)$$

$$+ c(X \cap Y, X - Y) + c(X - Y, Y - X)\}$$

$$- 2c(X - Y, Y - X)$$

$$= c(X) + c(Y) - 2c(X - Y, Y - X).$$

One can similarly verify the other equation. Q.E.D.

11.4.2. Feasibility

In this subsection we present an algorithm for deciding whether network N has multicommodity flows satisfying given demands, i.e., for testing feasibility. In what follows, we assume that $N = (G, P, c)$ belongs to class C_1.

We may assume that $b + 1 \leqslant 2k$. Otherwise, add new exterior edges of capacity zero as follows. Let the vertices on B that are source or sink be $v_{i1}, v_{i2}, \ldots, v_{ij}$, taken in clockwise order on B. The new edges are $(v_{i1}, v_{i2}), (v_{i2}, v_{i3}), \ldots, (v_{ij}, v_{i1})$. Clearly $|B'| = j \leqslant 2k$ if B' is the new outer boundary.

First we define some terms. For $X \subset V$, define the *margin* $m(X)$ of a cut $E(X)$ as $m(X) = c(X) - d(X)$. For $e, e' \in B$, define $m(e, e')$ as follows:

$$m(e, e') = \min\{m(X) \mid X \subset V, E(X) \cap B = \{e, e'\}\},$$

where $m(e, e') = \infty$ if there exists no $X \subset V$ such that $E(X) \cap B = \{e, e'\}$. That is, $m(e, e')$ is the minimum margin of cuts containing edges e and e' of B. It should be noted that e and e' are not always distinct in the definition above, and that $m(e, e) = \infty$ unless e is a bridge.

Consider the planar network N depicted in Fig. 11.2(a) to illustrate the terms above. If $X = \{u, v, w, s_2\}$ then $c(X) = 10$, $d(X) = 4$ and $m(X) = 6$. If $X = \{s_1, t_2\}$ then $m(X) = 0$. If $e = (s_1, v)$ and $e' = (s_2, s_3)$, then $m(e, e') = m(\{v, s_2\}) = 4$.

Combining Theorem 11.2 and Lemma 11.1, we obtain the following lemma.

Lemma 11.3. *A planar k-network $N = (G, P, c)$ belonging to class C_1 has k-commodity flows satisfying the demands if and only if $m(e, e') \geqslant 0$ for every $e, e' \in B$.*

Proof. We have to prove the implication in both directions. First assume that the network has k-commodity flows satisfying the demands. Then by Theorem 11.2 the network satisfies the cut condition. The cut condition says that $c(X) \geqslant d(X)$ for all X, so $m(X) \geqslant 0$ for all X, so $m(e, e') \geqslant 0$ for any e and e'.

To prove the converse assume that $m(e, e') \geqslant 0$ for all e and e' in B. This implies immediately that the cut condition is satisfied for any set X such that $|E(X) \cap B| = 1$ or 2. The cut condition is also satisfied for any set X such that $|E(X) \cap B| = 0$, because then the vertices of B are contained in X or $V - X$. Since all sources and sinks are on the boundary B, we have $d(X) = 0$, which implies the cut condition for these sets. Because of planarity, any set X that has the property that both $G - X$ and $G - (V - X)$ are connected satisfies $|E(X) \cap B| = 0, 1$, or 2. Therefore by Lemma 11.1 the network satisfies the cut condition for all X, and so by Theorem 11.2 there are k-commodity flows satisfying the demands.   Q.E.D.

To test the feasibility, one must check whether every $e, e' \in B$ satisfies $m(e, e') \geq 0$. For the purpose one must compute $c(X)$ and $d(X)$ for all $X \subset V$ with $|E(X) \cap B| = 1, 2$. Since there may exist an exponential number of these X's, the straightforward method cannot guarantee the polynomial-time bound.

Since all the sources and sinks are on the boundary of the outer face, clearly $d(X) = d(Y)$ whenever $E(X) \cap B = E(Y) \cap B$. Hence if $e, e' \in B$ are fixed then $d(X)$ is constant for all X with $E(X) \cap B = \{e, e'\}$. Denote the constant by $d(e, e')$, and let

$$c(e, e') = \min\{c(X) \mid X \subset V \text{ and } E(X) \cap B = \{e, e'\}\},$$

then

$$m(e, e') = c(e, e') - d(e, e').$$

Thus we shall compute $c(e, e')$ and $d(e, e')$ quickly.

One can compute $d(e, e')$ for a fixed $e \in B$ and all $e' \in B$ in $O(k + b)$ time. Remember that $|P| = k$ and $|B| = b + 1 \leq 2k$. These values can be easily updated for the new e next to the current e on B. Thus we can compute $d(e, e')$ for all $e, e' \in B$ in $O(k + b^2)$ time.

We now show how to compute $c(e, e')$. Construct a new graph $G^* = (V^*, E^*)$ as follows: replace every edge on B by a pair of multiple edges; construct the dual of the graph; and finally delete from it the vertex corresponding to the outer face of G. G^* is illustrated in Fig. 11.7 for G depicted in Fig. 11.2(a). [Precisely, G^* are not "graphs", but "multigraphs", that is, they may have multiple edges. However this does not affect the arguments in what follows.] Consider the capacity function c as a length function of G^*. Then the minimum value $c(e, e')$ of cuts containing e and e' is equal to the length of a shortest path joining two vertices corresponding to e and e' in G^*. Therefore we can compute $c(e, e')$ in $O(T(n))$ time for a fixed $e \in B$ and all $e' \in B$ by applying a single-source shortest path algorithm to G^*. Note that $|E^*| = O(n)$ and $|V^*| = O(n)$ since G is planar. Thus all $c(e, e')$ can be computed either in

$O(bT(n))$ time by repeating the computation above for each $e \in B$ or in $O(T_{all}(n))$ time by directly applying an all-pair shortest path algorithm to G^*.

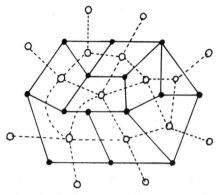

Fig. 11.7. G and G^*. (G is drawn in solid lines, and G^* in dotted lines.)

Above all the feasibility can be tested in $O(bT(n))$ or $O(T_{all}(n))$ time.

11.4.3. Algorithm DELTAFLOW

The details of algorithm MULTIFLOW1 will be given in the next subsection. In order to explain the core of MULTIFLOW1 we present, in this subsection, an "algorithm" DELTAFLOW which finds k-commodity flows in a planar network N, but does not run necessarily in polynomial time.

procedure DELTAFLOW(N);
begin
1. select an appropriate edge e_0 on the boundary B incident with a source or a sink (assume w.l.o.g. that $e_0 = (s_i, v)$);
2. determine an amount D as an appropriate positive number $\leqslant c(e_0)$;
3. push D units of flow f_i through e_0;
4. let $N' = (G', P', c')$ be a new $(k + 1)$-network such that
   $P' = P \cup \{(s_{k+1}, t_{k+1})\}$, where $s_{k+1} = v$ and $t_{k+1} = t_i$;
   $c'(e) = c(e)$ if $e \in E - e_0$, and $c'(e_0) = c(e_0) - D$; and
   $G' = (V, E')$ where $E' = E$ if $c'(e_0) > 0$, or $E' = E - e_0$ otherwise;
5. define the new demands $d'_j (1 \leqslant j \leqslant k + 1)$ as follows: $d'_i = d_i - D$, $d'_{k+1} = D$, and $d'_j = d_j$ if $1 \leqslant j \leqslant k$ and $j \neq i$;

6. apply DELTAFLOW recursively to the new network N' to find $(k + 1)$-commodity flows $f_1, f_2, \ldots, f_{k+1}$ of demands $d'_1, \ldots, d'_{k+1}$ in N';
7. superimpose three flows, f_i, f_{k+1} and the D units of f_i pushed through edge e_0, into a new single flow f_i, that is, define $f_i : E \to R$ as follows:

$$f_i(e) := f_i(e) + f_{k+1}(e) \qquad \text{if } e \in E - e_0,$$

$$f_i(e_0) := f_i(e_0) + f_{k+1}(e_0) \pm D \quad \text{if } e_0 \in E',$$

$$f_i(e_0) := \pm D \qquad\qquad\quad \text{if } e_0 \notin E',$$

where the sign $\pm$ depends on the orientation of e_0;
8. output $f_1, f_2, \ldots, f_k$ as the k-commodity flows in N;
end.

See Fig. 11.2 for an illustration. If two units of flow f_1 are pushed through e_0, then the 3-network in Fig. 11.2(a) results in a new 4-network in Fig. 11.2(b).

The new planar network N' needs to satisfy the cut condition for the new demands $d'_1, \ldots, d'_{k+1}$ so that algorithm DELTAFLOW works well. The next lemma gives the requirement for value D, where we define $m(e_0; Q)$ for a path Q on B as follows: $m(e_0; Q) = \min\{m(e_0, e) \mid e \in Q\}$ if Q is not empty; $m(e_0; Q) = \infty$ otherwise.

Lemma 11.4. (a) *Let N be a planar network satisfying the cut condition, and let edge $e_0 = (s_i, v) \in B$ be incident with a terminal s_i of demand d_i. Let Q be the path on B joining v and the other terminal t_i and not containing edge $e_0 = (v, s_i)$. If*

$$D = \min\{c(e_0), d_i, m(e_0; Q)/2\},$$

then network N' satisfies the cut condition for demands $d'_1, \ldots, d'_{k+1}$. (For N in Fig. 11.2(a) $D = \min\{3, 3, 4/2\} = 2$.)

(b) *For every $X \subset V$ the margin $m(X)$ does not increase during the execution of DELTAFLOW(N). In particular, once $m(X)$ becomes 0, it remains unchanged thereafter.*

Proof. Assume that e_0 is not deleted, that is, $E' = E$. (The proof for the other case is similar.) We shall show that $m(X) = c(X) - d(X)$ remains nonnegative for every $X \subset V$ such that $|E(X) \cap B| = 1, 2$. Clearly $c(X)$ or $d(X)$ changes the value only if $e_0 \in E(X)$. We can assume without loss of generality that $t_i \notin X$; otherwise consider $m(V - X)$. Denote by $m'(X)$ the margin in the new network N'. If $s_i \notin X$ and $v \in X$,

$$m'(X) = (c(X) - D) - (d(X) + D) = m(X) - 2D.$$

If $s_i \in X$ and $v \notin X$,

$$m'(X) = (c(X) - D) - (d(X) - D) = m(X).$$

Thus $m(X)$ never increases. It decreases (exactly by $2D$) only if there exists an edge $e \in Q$ such that $E(X) \cap B = \{e_0, e\}$. Hence we have $m'(X) \geq 0$ since $m(X) \geq 0$ and

$$D = \min\{c(e_0), d_i, m(e_0; Q)/2\}.$$

The first two terms of the right hand side above are necessary to guarantee that $c'(e_0)$ and d'_i are nonnegative. Q.E.D.

The following lemma asserts that G has at least one edge e_0 with positive D whenever N satisfies the cut condition.

Lemma 11.5. *Let N be a planar network satisfying the cut condition. If at least one demand is positive, then there exists $e_0 \in B$ such that $m(e_0, e) > 0$ for every $e \in Q$, where e_0 is incident clockwise with a terminal, say $e_0 = (s_i, v)$, and Q is the path on B going clockwise from v to t_i.*

Proof. Suppose that the claim is not true, then there exists $X \subset V$ such that $|E(X) \cap B| = 1, 2$ and $m(X) = c(X) - d(X) = 0$. Let X be such a minimal set. Assume that $|E(X) \cap B| = 2$. (The proof for the case that $|E(X) \cap B| = 1$ is similar.) Let $E(X) \cap B = \{(x, y), (z, w)\}$. We can assume without loss of generality that $y, z \in X$ and $x, w \notin X$, and that the vertices x, y, z, and w appear, in this order, clockwise on B. Since $c(X) > 0$, $d(X) > 0$. Therefore there exist source–sink pairs of positive demands with a source or sink in X and the other in $V - X$. Among these let (s_i, t_i) be a pair such that s_i or t_i (say t_i) first appears when traversing B clockwise from w. Let v be the vertex on B clockwise next to s_i, and let $e_0 = (s_i, v)$. Let Q be the path on B going clockwise from v to t_i. Since we assume that the claim is not true, there exists $Y \subset V$ such that $v \in Y$, $m(Y) = c(Y) - d(Y) = 0$ and $E(Y) \cap B = \{e_0, e\}$ for an edge $e \in Q$. By the choice of (s_i, t_i) above we have $d(X - Y, Y - X) = 0$. Hence by Lemma 11.2 we have

$$d(X \cap Y) + d(X \cup Y)$$

$$= d(X) + d(Y)$$

$$= c(X) + c(Y)$$

$$= c(X \cap Y) + c(X \cup Y) + 2c(X - Y, Y - X).$$

Since N satisfies the cut condition, $d(X \cap Y) \leqslant c(X \cap Y)$ and $d(X \cup Y) \leqslant c(X \cup Y)$. The above three equations imply that

$$d(X \cap Y) = c(X \cap Y), \quad d(X \cup Y) = c(X \cup Y),$$

and

$$c(X - Y, Y - X) = 0.$$

Therefore $E(X - Y, Y - X) = \varnothing$, so $X \cap Y \neq \varnothing$. Hence $\varnothing \neq X \cap Y \subset X - \{y\}$, and clearly $|E(X \cap Y) \cap B| = 2$. This contradicts the minimality of X.   Q.E.D.

We are now ready to prove Theorem 11.2.

Proof of Theorem 11.2. By rational approximation it suffices to deal with the case when the capacities and demands are all rational valued; and hence it suffices to prove it when the capacities and demands are integer-valued (by multiplying throughout by a common denominator). We shall in fact prove a stronger result: that if the capacities and demands are integer-valued, then the multicommodity flows may be chosen half-integer-valued; or equivalently, if the capacities and demands are even-integer-valued, then the multicommodity flows may be chosen integer-valued.

Indeed, we shall prove that algorithm DELTAFLOW with choosing D as in Lemma 11.4(a) finds integral multicommodity flows under a weak hypothesis: the capacities and demands are integer-valued and that $m(X)$ is even for each $X \subset V$. The proof is by induction on the sum of the capacities over all the edges. Assume that the claim is true for all the networks such that the sum is less than a particular value ($\geqslant 1$), and assume that a network N has the sum equal to the value. Lemma 11.5 implies that if at least one demand is positive then N contains an edge $e_0 = (s_i, v)$ on B such that $m(e_0, e) > 0$ for every edge e on the path Q on B connecting v and t_i. Hence Lemma 11.4 implies that one can choose a positive integer D and the network N' in the 4th line of DELTAFLOW satisfies both the cut condition and the above hypothesis. Note that $m(X)$ is even for each $X \subset V$ in N'. Since the sum of capacities of N' is less than that of N, DELTAFLOW correctly finds integral $(k + 1)$-commodity flows in N' and hence the k-commodity flows superimposed at the 7th line are correctly integral k-commodity flows in N.   Q.E.D.

Clearly DELTAFLOW terminates finitely if the capacities and demands are all integers. However this is not the case if they are real-valued. There are three obstructions to the correctness of polynomial boundedness of DELTAFLOW:

(1) If edge e_0 is selected arbitrarily, DELTAFLOW does not always terminate in polynomial time: when there exists an edge e of infinite capacity on the boundary B, DELTAFLOW possibly lets a flow go and return infinitely many times through e;

(2) The number of source–sink pairs increases, and the representation of flow functions would require a great deal of space;

(3) The new graph G' may be disconnected even if G is connected.

11.4.4. Algorithm MULTIFLOW1

In this subsection we present an algorithm MULTIFLOW1 for finding multicommodity flows in planar networks, which spends $O(kn + nT(n))$ time and $O(kn)$ space. MULTIFLOW1 uses the operation of DELTAFLOW, but is improved on the three obstructions mentioned at the end of the preceding subsection in the following way.

Obstruction (1). In order to make the algorithm run in polynomial time, we select both edge e_0 and flow f_i which is pushed through e_0, as follows. First select an arbitrary edge, say $e_0 = (v_0, v_1)$, on B; apply the operation of DELTAFLOW for e_0 with respect to each flow having v_0 as a source (or sink), in the order that the corresponding sink (or source) appears on B in clockwise order. For details, see procedure PUSH(N, e_0) given later in this subsection. Next select the edge *clockwise* next to e_0 on B as the new e_0, and apply the same procedure as above for the new e_0. Repeat this procedure until there exists no source–sink pair. We will show later in Section 6 that MULTIFLOW1 terminates before e_0 traverses all the edges once in each of its two directions. Thus we can guarantee the polynomial-time boundedness of MULTIFLOW1.

Obstruction (2). Although MULTIFLOW1 also makes a new source–sink pair (s_{k+1}, t_{k+1}), it does not introduce the new flow function f_{k+1}, but simply attaches to the pair a number indicating the kind of commodity between s_{k+1} and t_{k+1}. As shown later in the succeeding subsection, the number of source–sink pairs is at most $O(k + n)$ throughout the execution of MULTIFLOW1.

Obstruction (3). If graph G' is disconnected by the deletion of edge e_0, then we find multicommodity flows in each connected component of G'. Note that G' consists of two components.

We are now ready to present algorithm MULTIFLOW1 which improves on the three obstructions.

procedure MULTIFLOW1(N);
begin
   if there exist $e, e' \in B$ with $m(e, e') < 0$
      then print "No multiflows of demands" and stop;
   for each $e \in E$ and i $(1 \leqslant i \leqslant k)$ *do* $f_i(e) := 0;$ {initialization}
   $p := k;$ {the number of source–sink pairs}
   $e_0 :=$ an arbitrary edge (v_0, v_1) on B;
   ROTATE(N, e_0)
end.

procedure ROTATE(N, e_0);
begin
   {this procedure rotates e_0 around B in clockwise order}
   delete all source–sink pairs (s_i, t_i) such that $s_i = t_i$ or $d_i = 0$;
   if network N has a source–sink pair *then*
     begin
       PUSH(N, e_0);
       {procedure PUSH(N, e_0) pushes flows through $e_0 = (v_0, v_1)$ by applying
       the operation of DELTAFLOW to each flow having v_0 as a source, in the
       order that the corresponding sink appears on B in clockwise order. Note
       that $c(e_0)$ may decrease when the procedure terminates.}
       if $c(e_0) > 0$ *then*
         begin
           $e_0 :=$ the edge clockwise next to e_0 on B;
           ROTATE(N, e_0)
         end
       else {e_0 is saturated, i.e., $c(e_0) = 0$.}
         begin
           $G := G - e_0;$ {delete e_0 from G}
           if G is connected *then*
              begin
                let e_0 be the edge on the new boundary B of G, joining v_0 and
                the clockwise next vertex on B;
                ROTATE(N, e_0)
              end
           else for each connected component G_j $(j = 1, 2)$
              of G *do*
              begin
                let N_j be the subnetwork of N with graph G_j;
                let e_0 be the edge on the boundary of the outer face B_j of G_j,
                joining v_0 or v_1 and the next clockwise vertex on B_j;

ROTATE(N_j, e_0)
   end
  end
 end
end;

Before presenting procedure PUSH(N, e_0), consider how to decide which flows and what amounts can be pushed through e_0. Let $e_0 = (v_0, v_1)$. Suppose that terminals $s_1, s_2, \ldots, s_l$ are assigned to v_0, and that the corresponding terminals are $t_1, t_2, \ldots, t_l$, taken in clockwise order from v_1. Let Q_1 be the path on B clockwise going from v_1 to t_1, and Q_i the path from t_{i-1} to t_i ($2 \leqslant i \leqslant l$). (See Fig. 11.8.) Note that Q_i is possibly empty. Given $m(e_0, e')$ for all $e' \in B$, one can compute $m(e_0; Q_1)$, and so decide the amount D_1 of the flow between s_1 and t_1 to be pushed through e_0, in $O(|Q_1| + 1)$ time (see Lemma 11.4(a)). When D_1 units of the flow are pushed through e_0, $m(e_0, e')$ decreases by $2D_1$ if $e' \in Q_1$, and remains unchanged if $e' \in Q_2 \cup \cdots \cup Q_l$. (See the proof of Lemma 11.4.) In order to decide amount D_2 of the flow between s_2 and t_2 to be pushed through e_0, a trivial algorithm alters the values $m(e_0, e')$ for $e' \in Q_1$ and finds the minimum of $m(e_0, e')$ over all $e' \in Q_1 \cup Q_2$, so it spends $O(|Q_1| + |Q_2| + 1)$ time. Thus a straightforward algorithm, repeating this procedure, would require $O(n^2)$ time to decide all D_i. However, by updating the value $x(i) = \min\{c(e_0), m(e_0, e')/2 \,|\, e' \in Q_1 \cup \cdots \cup Q_i\}$ from $x(i-1)$, one can decide each D_i, $1 \leqslant i \leqslant l$, in only $O(|Q_i| + 1)$ time as shown in PUSH(N, e_0) below. Thus, given $m(e_0, e')$ for all $e' \in B$, one can decide all D_i, $1 \leqslant i \leqslant l$, in $O(n + l)$ time, since

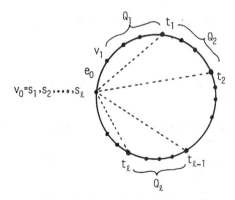

Fig. 11.8. Illustration for $Q_1, Q_2, \ldots, Q_l$.

$$\sum_{i=1,...,l} (|Q_i| + 1) \leqslant |B| + l = O(n + l).$$

procedure PUSH(N, e_0);
begin
   {this procedure pushes flows through edge $e_0 = (v_0, v_1)$}
   $x := \min\{c(e_0), m(e_0; Q_1)/2\}$; {$x = x(1)$}
   $i := 1$
   while ($x > 0$) *and* ($i \leqslant l$) *do*
     if $d_i \leqslant x$ *then* {the flow between s_i and t_i can be entirely pushed through e_0}
       begin
          $D_i := d_i$;
          $s_i := v_1$; {surrogate source}
          $c(e_0) := c(e_0) - D_i$; {residual capacity}
          {update $x(i + 1)$ from $x(i)$}
          if $i = l$ *then* $x := 0$
             else $x := \min\{x - d_i, m(e_0; Q_{i+1})/2\}$;
          $i := i + 1$ {clockwise next pair}
       end;
     else {the flow between s_i and t_i can be partly pushed through e_0}
       begin
          $D_i := x$;
          $d_i := d_i - D_i$; {residual demand}
          $c(e_0) := c(e_0) - D_i$;
          create a new source–sink pair (s_{p+1}, t_{p+1}) of the same commodity as
          that between s_i and t_i where $s_{p+1} = v_1$ and $t_{p+1} = t_i$;
          $d_{p+1} := D_i$; {demand of new pair}
          $p := p + 1$; {the number of pairs increases by one}
          $x := 0$;
          $i := i + 1$
       end;
   $j := i - 1$; {the first j flows can be pushed through edge e_0}
   for $i := 1$ *to* j *do*
     begin
       let (s_i, t_i) be a source–sink pair for commodity r, where $1 \leqslant r \leqslant k$;
       $f_r(e_0) := f_r(e_0) \pm D_i$;
       {the sign $\pm$ depends on both the orientation of edge e_0 and whether v_0 is
       source s_i or sink t_i}
     end;
end;

11.4.5. *Time and space of MULTIFLOW 1*

Since MULTIFLOW1 is essentially the same as DELTAFLOW, MULTI-FLOW1 correctly finds multiflows whenever it terminates. In the remainder of this subsection we discuss the time and space complexity of algorithm MULTIFLOW1.

Some lemmas

When procedure PUSH(N, e_0) is executed with $e_0 = (v_0, v_1)$, one of the following three cases occurs depending on which term attains the minimum in the equation $D = \min\{c(e_0), d_i, m(e_0, Q)/2\}$ (see Lemma 11.4(a)):

Case (1): edge e_0 becomes saturated and will be deleted by ROTATE;

Case (2): the $s_i - t_i$ flow of demand d_i is entirely pushed through e_0, so source s_i disappears from v_0 and a "surrogate source" s_i is constructed at v_1; and

Case (3): a "bottleneck" edge e is known to exist on B somewhere between v_1 and t_i, that is, there exists $X \subset V$ such that $m(X) = 0$, $v_1 \in X$, and $E(X) \cap B = \{e_0, e\}$.

Similarly as Lemma 11.5 one can prove the following lemma.

Lemma 11.6. *Suppose that there exists $X \subset V$ such that $m(X) = 0$, $v_0 \in X$, and $E(X) \cap B = \{(v_b, v_0), (v_i, v_{i+1})\}$ ($i < b$). Denote by R the path on B going from v_0 to v_{i+1} in clockwise order. If procedure MULTIFLOW1(N) is executed with first assigning edge (v_0, v_1) to variable e_0, then at least one edge on R is deleted on a first traversal of R.*

Proof. Assume that there is a network N for which the lemma is not true for some $X \subset V$, that is, Case (1) never occurs on a first traversal of R. Moreover assume that X is minimum in cardinality among all these X's. Suppose that Case (3) never occurs. Then there would be no source or sink of a positive demand in X, so $d(X) = 0$ when PUSH($N, (v_i, v_{i+1})$) terminates. By Lemma 11.4(b) $m(X) = 0$ at that time since $m(X) = 0$ when procedure MULTIFLOW1(N) started. However $c(X) > 0$ since no edge on R is deleted. This is a contradiction. Thus we have shown that Case (3) occurs for an edge on R.

Assume that $e = (v_j, v_{j+1})$ is the first one of such edges on R, where $0 \leq j \leq i$. Then all the vertices $v_0, \ldots, v_{j-1}$ are not sources or sinks of positive demands, and there exists $Y \subset V$ such that $m(Y) = 0$, $v_{j+1} \in Y$ and $E(Y) \cap B = \{e, e'\}$ for some $e' = (v_l, v_{l+1}) \in B$ with $l < b$. Furthermore we may assume that Y contains no sources or sinks corresponding to sinks or sources assigned to v_j. Therefore $d(X - Y, Y - X) = 0$. Combining this equation with $m(X) = m(Y) = 0$ through Lemma 11.2, we have

$$m(X \cap Y) + m(X \cup Y) = -2c(X - Y, Y - X) \leqslant 0.$$

On the other hand Lemma 11.4(a) implies that the current network N satisfies the cut condition, and hence $m(X \cap Y)$, $m(X \cup Y) \geqslant 0$. If $j = i$, then

$$c(X - Y, Y - X) \geqslant c((v_i, v_{i+1})) > 0,$$

contradicting the equation above. Thus $j < i$. Then $\varnothing \neq X \cap Y \subset X - \{v_j\}$ and $m(X \cap Y) = 0$. Therefore the assumption on the minimality of X implies that at least one edge is deleted on the first traversal of the path on B clockwise going from v_{j+1} to the first vertex not in $X \cap Y$. Clearly this edge is on R, a contradiction. Q.E.D.

Lemma 11.7. *Algorithm MULTIFLOW1(N) assigns no single edge to the variable e_0 more than once for each of its two orientations.*

Proof. Assume that $N = (G, P, c)$ is a network for which the lemma is not true, and that G has a minimum number of edges among such networks; clearly the number is positive. We may assume without loss of generality that edge (v_0, v_1) is first assigned to e_0.

We now show that at least one edge, other than the last edge (v_b, v_0), is deleted on a first traversal of B. Suppose on the contrary that no edge is deleted. If Case (3) never happens for any edge in $B - \{(v_b, v_0)\}$, then MULTIFLOW1(N) would terminate before the first traversal of $B - \{(v_b, v_0)\}$ has been completed, contrary to the assumption. Thus Case (3) occurs for an edge $e = (v_j, v_{j+1})$ on $B - \{(v_b, v_0)\}$. Then there exists $X \subset V$ such that $m(X) = 0$, $v_{j+1} \in X$ and $E(X) \cap B = \{(v_j, v_{j+1}), (v_l, v_{l+1})\}$ ($l < b$). Denote by R the path on B going from v_{j+1} to v_{l+1} in clockwise order. By Lemma 11.6 at least one edge must be deleted on the first traversal of path R. This contradicts the supposition.

Let $e = (v_i, v_{i+1}) \in B$ be the first edge deleted from the graph, where $i < b$. Note that MULTIFLOW1(N) has assigned to e_0 each of the edges $(v_0, v_1), \ldots, (v_i, v_{i+1})$ once so far. Assume that network N results in $N' = (G', P', c')$ when procedure PUSH(N, e) finishes, where $G' = G - e$. Then the following two cases happen.

Case 1: G' is connected. Reapply procedure MULTIFLOW1 to the new network $M = N'$ by first assigning (v_0, v_1) to e_0, and consider the behavior by dividing the time period into two parts: (a) while e_0 is $(v_0, v_1), \ldots,$ or (v_{i-1}, v_i); (b) after e_0 becomes (v_i, u), where u is the vertex clockwise next to v_i on the new outer boundary of G'.

Consider the period (a). By Lemma 11.4(b) margins have never increased in

MULTIFLOW1(N). Since $e \neq (v_b, v_0)$, no "surrogate" source or sink has been constructed at v_0 in MULTIFLOW1(N). Therefore no flow can be pushed through edges $(v_0, v_1), \ldots, (v_{i-1}, v_i)$ in MULTIFLOW1(M). Therefore the network M is not altered at all during the period (a), that is, $M = N'$ when e_0 becomes (v_i, u).

Thus during (b) the behavior of MULTIFLOW1(M) is identical to that of MULTIFLOW1(N). Since the number of edges of G' is one less than that of G, the assumption of the minimality of G implies that MULTIFLOW1(M) assigns no single edge of G' to e_0 more than once for each of its orientations. Hence MULTIFLOW1(N) assigns no single edge of G to variable e_0 more than once for each of its orientations, contrary to the assumption.

Case 2: G' is disconnected. Let G_1 and G_2 be the two connected components in G'. We may assume that G_1 contains v_i, and G_2 contains v_{i+1}. Let $N_i = (G_i, P_i, c_i)$ be the resulting networks ($i = 1, 2$). The behavior of procedure MULTIFLOW1(N) is identical with a combination of two behaviors: the behavior of MULTIFLOW1(N_1) beginning at (v_0, v_1); and the behavior of MULTIFLOW1(N_2) beginning at (v_{i+1}, v_{i+2}). Since both G_1 and G_2 have fewer edges than G, MULTIFLOW1(N_1) and MULTIFLOW1(N_2) assign no single edge of G_1 or G_2 to e_0 more than once for each of its orientations, contrary to the assumption. Q.E.D.

Lemma 11.7 guarantees the polynomial boundedness and so termination of MULTIFLOW1.

Fig. 11.9 illustrates a partial traversal of variable e_0 in the network N of Fig. 11.2(a). The deleted edges are drawn in dashed lines. A number i and an arrow next to an edge indicate that MULTIFLOW1(N) assigns the edge to e_0 for the orientation of the arrow in the ith execution of ROTATE (and PUSH). An edge has been assigned to e_0 once for each of the two orientations.

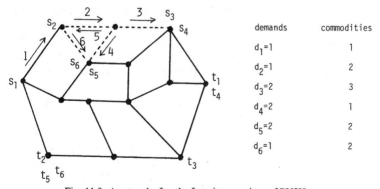

demands	commodities
$d_1 = 1$	1
$d_2 = 1$	2
$d_3 = 2$	3
$d_4 = 2$	1
$d_5 = 2$	2
$d_6 = 1$	2

Fig. 11.9. A network after the first six executions of PUSH.

Data structure and space. A graph G is represented by the adjacency lists, in each of which the edges adjacent to a vertex are stored in the order of the planar embedding, clockwise around the vertex. Thus, given an edge e, the edge clockwise next to e around an end of e can be directly accessed. Since G is planar, $|E| = O(n)$. Therefore G can be represented in $O(n)$ space.

Each of the k flow functions $f_1, f_2, \ldots, f_k$ is represented by an array of size $|E|$. Thus the representation of k-commodity flows uses $O(kn)$ space.

The set P of source–sink pairs is represented by a multigraph $G_P = (V(B), P)$, each edge of which corresponds to a source–sink pair. Two numbers are associated with each pair (s_i, t_i): a real-valued demand d_i; and an integer $r(i)$ $(1 \leqslant r(i) \leqslant k)$ indicating the kind of commodity between s_i and t_i. As shown in Lemma 11.8 below, the number of source–sink pairs is at most $O(k + n)$ throughout the execution of MULTIFLOW1. Thus G_P is represented in $O(k + n)$ space. Hence MULTIFLOW1 uses $O(kn)$ space in total.

Lemma 11.8. *The number of source–sink pairs is at most* $O(k + n)$ *throughout the execution of MULTIFLOW 1.*

Proof. Consider procedure PUSH(N, e_0). Let j be the integer decided in PUSH(N, e_0). For every i, $1 \leqslant i \leqslant j - 1$, s_i is moved to v_1. Only for j may a new source–sink pair be created. Therefore one execution of PUSH(N, e_0) increases the number of source–sink pairs by at most one. Lemma 11.7 implies that MULTIFLOW1 calls PUSH(N, e_0) at most $2|E|$ times in total. Thus we have established the claim.   Q.E.D.

Computation time. Consider the computation time for one execution of procedure PUSH(N, e_0). We must compute $m(e_0, e')$ for all $e' \in B$. As shown in subsection 11.4.2, one can compute them in $O(k + T(n))$ time by applying a shortest path algorithms to G^*. We have shown in the preceding subsection that, given $m(e_0, e')$ for all $e' \in B$, one can decide all D_i, $1 \leqslant i \leqslant j$, in $O(n + l)$ time. By Lemma 11.8 l is at most $O(k + n)$. Clearly PUSH(N, e_0) updates $c(e_0)$ and $f_r(e_0)$ in $O(l) = O(k + n)$ time. Hence we have shown that one execution of PUSH(N, e_0) can be done in $O(k + T(n))$ time.

Since MULTIFLOW1 calls PUSH(N, e_0) $O(n)$ times by Lemma 11.7, PUSH(N, e_0) spends $O(kn + nT(n))$ time in total. MULTIFLOW1(N) decides the feasibility in $O(bT(n))$ time as shown in subsection 11.4.2, and the *for* statement in MULTIFLOW1 spends $O(kn)$ time. The remaining time is for procedure ROTATE(N, e_0). One execution of ROTATE(N, e_0) can be done in at most $O(n)$ time, exclusive of the time spent by PUSH(N, e_0) called there. Since ROTATE(N, e_0) is called $O(n)$ times by Lemma 11.7, it spends $O(n^2)$

time in total. Thus we have shown that Algorithm MULTIFLOW1 spends $O(kn + nT(n))$ time in total.

We now have the following theorem.

Theorem 11.3. *Algorithm MULTIFLOW1 correctly finds multicommodity flows of given demands in a planar network $N = (G, P, c)$ if all the sources and sinks are on the boundary of the outer face of a planar graph G. It spends $O(kn + nT(n))$ time and $O(kn)$ space if there are n vertices and k source–sink pairs.*

It is interesting that the values of the obtained flows are half integers if the capacities and demands are all integers. Neither a claim similar to Theorem 11.2 holds for planar *directed* graphs, nor the algorithm correctly works for them, even in the case of two-commodity flows. Fig. 11.10 depicts a planar directed graph in which all sources and sinks lie on the outer boundary. Although the graph satisfies the cut condition of the directed version, it has no two-commodity flows realizing the given demands.

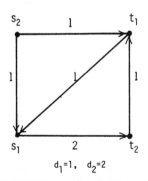

Fig. 11.10. A planar directed graph. (The number associated with each edge represents the capacity of the edge.)

Okamura [Oka83] has generalized Theorem 11.2 to the class C_{12} of planar graphs in which all the terminals are located on two specified face boundaries, while Suzuki, Nishizeki and Saito [SNS85] have given an algorithm for the case similar to MULTIFLOW1.

11.5. Multicommodity flows for C_a

In this section we deal with the multicommodity flow problem on another class of planar undirected graphs. Join all the sources of a given planar graph G to the corresponding sinks. Denote the resulting graph by G_a; one example is depicted in Fig. 11.1(d). We deal with the class C_a of planar graphs G having planar G_a. Seymour [Sey81] has proved that the Max Flow–Min Cut theorem holds for such planar graphs, while Matsumoto, Nishizeki and Saito [MNS86] have given an algorithm. The algorithm reduces the multicommodity flow problem on such planar graphs to the maximum weighted matching problem, which is known to be polynomial-time solvable. More precisely, testing the feasibility of mutlticommodity flows in G is reduced to detecting a negative cycle in the dual G_a^* of G_a, whereas finding the multicommodity flows in G is reduced to finding, $O(n)$ times, a minimum cycle passing through a specified vertex in G_a^* or its variants. Then detecting the negative and minimum cycles in a graph is reduced to finding a maximum weight matching in a certain graph constructed for the graph. Thus the feasibility can be tested by solving, once, the weighted matching problem, and the multicommodity flows of given demands can be found by solving the matching problem $O(n)$ times. There is an $O(n^{3/2} \log n)$ algorithm which finds a maximum matching in the graphs which appeared in our reductions and hence detects the negative and minimum cycles in any planar undirected graph. It is based on both Lipton and Tarjan's planar separator theorem (Theorem 9.4) and Galil, Micali and Gabow's matching algorithm [GMG82]. Consequently the feasibility can be tested in $O(n^{3/2} \log n)$ time and the multicommodity flows can be found in $O(n^{5/2} \log n)$ time.

11.5.1. Feasibility

Seymour [Sey81] gave the following theorem.

Theorem 11.4. *A network N belonging to class C_a has multicommodity flows if and only if N satisfies the cut condition.*

In this subsection we present an account of how to quickly check the feasibility using Theorem 11.4.

Denote by G_a the graph obtained from a given planar graph $G = (V, E)$ by adding a new edge $e_{ai} = (s_i, t_i)$ to G for each i, $1 \leqslant i \leqslant k$, as depicted in Fig. 11.11. We call e_{ai} a *demand* or *negative* edge. (This terminology will be justified below.) Since G_a is planar, $k = O(n)$.

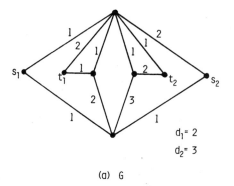

(a) G

$d_1 = 2$
$d_2 = 3$

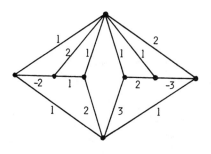

(b) G_a

Fig. 11.11. Planar graphs (a) G and (b) G_a.

Define a new capacity function $c_a : E_a \to R$ for $G_a = (V, E_a)$ as follows:

$$c_a(e) = \begin{cases} c(e) & \text{if } e \in E; \\ -d_i & \text{if } e = e_{ai}. \end{cases}$$

Thus only e_{ai}, $1 \le i \le k$, have negative capacities in G_a. Also define $E_a(X)$ and $c_a(X)$ in a way similar to $E(X)$ and $c(X)$, respectively. Then clearly $c_a(X) = m(X)$. Consider the dual G_a^* of the planar graph G_a. [Precisely G_a and G_a^* are "multigraphs". However this does not affect the arguments in what follows.] The graph in Fig. 11.12(a) is the dual of G_a in Fig. 11.11(b). We denote by e^* the edge of G_a^* corresponding to an edge e of G_a. We interpret c_a as a weight function of G_a^*. If $E(X)$ is a cutset of G, then $E_a(X)$ is a cutset of G_a and hence corresponds to a cycle of G_a^*. Thus Theorem 11.4 and Lemma 11.1 imply that network N has multicommodity flows of given demands if and only if G_a^* has no negative cycle. Hence the feasibility can be tested simply by checking the existence of negative cycles in G_a^*.

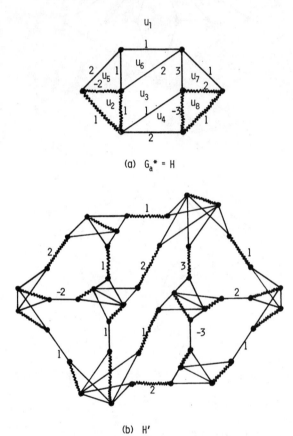

(a) $G_a^* = H$

(b) H'

Fig. 11.12. Graphs $H = G_a^*$ and H'. (The edges of M in H' and the edges of $F - M$ in H are drawn by wavy lines.)

Remember that a matching $M \subset F$ of a graph $H = G_a^* = (U, F)$ is a set of pairwise nonadjacent edges of H. If $|M| = |U|/2$, then M is called a *complete matching* of H. We now show that detecting a negative cycle in H can be reduced to finding a maximum weight matching in a certain graph directly constructed from H. Replace each vertex $v \in U$ of H by a "star" [Tut54] depicted in Fig. 11.13. Let $H' = (U', F')$ be the resulting graph. (See Fig.

11.12.) Define the edge weights of H' as follows: the surrogate of each edge $e \in F$ has the same weight as e; and all new edges in stars have zero weights. We have the following lemma.

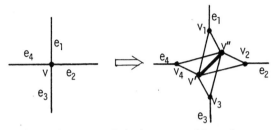

Fig. 11.13. The star substituting a vertex of degree four.

Lemma 11.9. *Let $M \subset F'$ be a maximum complete matching of $H' = (U', F')$. Then $H = (U, F)$ has a negative cycle if and only if the weight of $F - M$ is negative.*

Proof. Let $M \subset F'$ be any complete matching of H'. If M contains edge (v', v'') in a star substituting a vertex v of H, then M must contain all the edges incident to v in H and hence v has degree 0 in the subgraph of H induced by $F - M$. [Edge (v', v'') is drawn by a thick line in Fig. 11.13.] On the other hand, if M does not contain (v', v''), then v has degree 2 in the subgraph. Thus $F - M$ is a vertex-disjoint union of cycles in H. Furthermore one can easily observe from the construction of H' that the converse is also true: if $C \in F$ is a vertex-disjoint union of cycles in H then there exists a complete matching M of H' such that $C = F - M$ (see Fig. 11.12). Clearly the weight of $F - M$ is minimum if and only if the weight of M is maximum. The claim follows immediately from these facts.   Q.E.D.

If H'' is a graph the same as H' except for each weight being added the same large constant, then a maximum weight matching of H'' is a maximum weight complete matching of H' [Law76]. A maximum weight matching can be found in $O(|E||V|\log|V|)$ time for a general graph having $|V|$ vertices and $|E|$ edges by Galil, Micali and Gabow's algorithm [GMG82]. (See [Gal83] for the exposition of their algorithm and [BD83] for an alternative algorithm.) Since $|U'| = O(|F|)$ and $|F'| = O(|F|)$, a negative cycle in H can be detected in $O(|F|^2 \log|F|)$ time. Since G and H are planar, both $|F|$ and $|U|$ are $O(n)$, and hence a negative cycle in H can be detected in $O(n^2 \log n)$ time. Matsumoto, Nishizeki and Saito further improved the bound above to $O(n^{3/2} \log n)$ by using the divide-and-conquer. The key idea is that the class of graphs H' constructed from planar graphs H has an $n^{1/2}$-separator (see Chapter 9). They also gave another $O(kn \log n + k^3)$ algorithm by using Tobin's algorithm for detecting a nonpositive cycle [Tob75].

Thus we have the following theorem.

Theorem 11.5. *If a planar network N belongs to class C_a, then the feasibility of the k-commodity flows in N can be tested in* $O(n^{3/2} \log n)$ *or* $O(kn \log n + k^3)$ *time.*

11.5.2. Algorithm MULTIFLOW2

In this subsection we first give an algorithm MULTIFLOW2 which finds multicommodity flows in planar networks satisfying the cut condition. Then we verify the correctness. The algorithm is a generalization of the uppermost path algorithm in Section 11.2.

Algorithm. Each demand edge e_{ai}, $1 \leqslant i \leqslant k$, adjoins two faces F_{i1} and F_{i2} of the planar graph G_a. Let Q_{i1} be the path joining s_i and t_i on the boundary of F_{i1} without passing through e_{ai}. Similarly define Q_{i2} with respect to F_{i2}. (See Fig. 11.14.) MULTIFLOW2 is outlined as follows. Choose an appropriate path Q_{ij} among $2k$ ones; push appropriate units D of flow f_i through Q_{ij}; reduce by D the demand d_i and the capacities of edges in Q_{ij}; and delete saturated edges if any. Repeat this operation until k-commodity flows of given demands are obtained.

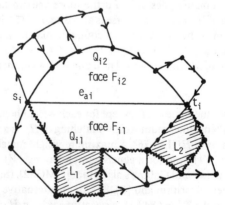

Fig. 11.14. Illustration of route R_i of flow f_i. (R_i is drawn by wavy lines. An arrow of an edge indicates the direction of f_i through the edge. No flow f_i passes through edges having no arrows.)

Of course, D can exceed neither d_i nor $c(e)$ for every $e \in Q_{ij}$. However, we wish to choose D as large as the resulting network does not violate the cut condition. (It is the crucial part of the algorithm.) Thus we determine D as follows:

$$D = \min\{-c_a(e_{ai}), \min\{c_a(e) : e \in Q_{ij}\}, c_a(Q_{ij})/2\}. \tag{11.1}$$

Here $c_a(Q_{ij})$ is the minimum capacity of cutsets of G_a containing exactly two edges of Q_{ij}. That is, if $c_a(e, e')$ is defined for two edges e and e' of Q_{ij} as follows:

$$c_a(e, e') = \min\{c_a(X) : E_a(X) \text{ is a cutset of } G_a$$

$$\text{and } E_a(X) \cap Q_{ij} = \{e, e'\}\}, \tag{11.2}$$

then

$$c_a(Q_{ij}) = \min\{c_a(e, e') : e, e' \in Q_{ij}\}.$$

We define $c_a(Q_{ij}) = \infty$ if $|Q_{ij}| = 1$, i.e., Q_{ij} is a single edge joining s_i and t_i.

We are now ready to present MULTIFLOW2.

procedure MULTIFLOW2;
begin
  for each $e \in E$ and i $(1 \le i \le k)$ *do* $f_i(e) := 0$; {initialization}
  while G_a has a demand edge {there remains an unsatisfied flow} *do*
    begin
      choose path Q_{ij} with positive D;
      {$1 \le i \le k$. $j = 1, 2$. Q_{ij} contains no demand edges.}
      {push D units of flow f_i through Q_{ij}}
      for each $e \in Q_{ij}$ *do*
        begin
          $f_i(e) := f_i(e) \pm D$;
          {the sign $\pm$ depends on both the orientation of edge e and the direction of path Q_{ij}}
          $c_a(e) := c_a(e) - D$; {residual capacity}
          if $c_a(e) = 0$ {e is saturated}
            then $G_a := G_a - e$ {delete e}
        end
      $c_a(e_{ai}) := c_a(e_{ai}) + D$;
      {decrease $d_i = -c_a(e_{ai})$ by D}
      if $c_a(e_{ai}) = 0$ {flow f_i has been satisfied}
        then $G_a := G_a - e_{ai}$ {delete demand edge e_{ai}}
    end
end.

Correctness. The following two lemmas must hold if algorithm MULTI-FLOW2 above correctly finds multicommodity flows.

Lemma 11.10. *If the original network $N = (G, P, c)$ satisfies the cut condition, then the new network $N' = (G', P', c')$ also satisfies the cut condition when D units of flow f_i have been pushed through Q_{ij}.*

Proof. We may assume that neither e_{ai} nor edges of path Q_{ij} are deleted and hence $G'_a = G_a$. (The proof for the other case is almost similar.) Let $c'_a(e)$ be the capacity of edge e in G'_a. Then $c'_a(e) = c_a(e) - D \,(> 0)$ for every edge e on Q_{ij}, and $c'_a(e_{ai}) = c'_a(e_{ai}) + D \,(< 0)$. Let $E_a(X)$ be any cutset of G'_a. Since $Q_{ij} \cup \{e_{ai}\}$ is a boundary of a face, we have

$$|E_a(X) \cap (Q_{ij} \cup \{e_{ai}\})| = 0 \text{ or } 2.$$

Therefore

$$c'_a(X) = \begin{cases} c_a(X) - 2D & \text{if } |E_a(X) \cap Q_{ij}| = 2; \\ c_a(X) & \text{otherwise.} \end{cases}$$

Since D satisfies Eq. (11.1), clearly $D \leqslant c_a(X)/2$ if $|E_a(X) \cap Q_{ij}| = 2$. Thus we have $c'_a(X) \geqslant 0$.   Q.E.D.

Lemma 11.11. *If network $N = (G, P, c)$ satisfies the cut condition, then G_a has at least one Q_{ij} with positive D among 2k paths.*

Proof. Assume on the contrary that there is no Q_{ij} with positive D, that is, any units of flow can be pushed through none of $2k$ paths. Then each Q_{ij} satisfies one of the following:

(a) Q_{ij} has a negative edge, that is, a demand edge;

(b) Q_{ij} has two "blocking" edges e and e' with $c_a(e, e') = 0$.

Since N satisfies the cut condition, Theorem 11.4 implies that N has a set of k-commodity flows satisfying given demands. For each Q_{ij} of $2k$ paths we define a route (path) R_i of flow f_i as follows. As shown in Fig. 11.14, let R_i be a path of G_a joining s_i and t_i such that positive units of flow f_i pass through R_i and the region bounded by $R_i \cup Q_{ij}$ contains a minimum number of faces in the interior. The assumption implies that the region contains at least one face.

Among $2k$ paths and all possible sets of k-commodity flows, we now choose a path Q_{ij} and a set of k-commodity flows for which region $R_i \cup Q_{ij}$ contains a minimum number of faces. If region $R_i \cup Q_{ij}$ consists of nonempty connected components $L_1, L_2, \ldots, L_h$, then choose any one of them, say L_b, $1 \leqslant b \leqslant h$.

(In the case of Fig. 11.14 $h = 2$.) The planarity of G_a implies that if a terminal lies in the proper interior of L_b then the other terminal lies in the interior of L_b. Clearly Q_{ij} and the boundary of L_b share a common edge such that either it is a demand edge or flow f_r other than f_i passes through it. Then source–sink pair (s_r, t_r) lies in L_b: otherwise, by interchanging routes of f_r and f_i as shown in Fig. 11.15, we can construct another set of k-commodity flows for which region $R_i \cup Q_{ij}$ contains fewer faces, a contradiction. Furthermore we may assume that route R_r of flow f_r is contained in L_b: otherwise, i.e., if R_r intersects with R_i, then one can construct another set of k-commodity flows for which R_r is contained in L_b, as shown in Fig. 11.16. However in this case region $R_r \cup Q_{rj'}$ ($j' = 1$ or 2) contains fewer faces than L_b, contrary to the assumption. Q.E.D.

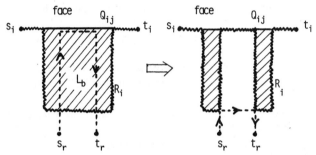

Fig. 11.15. Alternation of a route of flow f_i. (A route of f_r is drawn by dashed lines.)

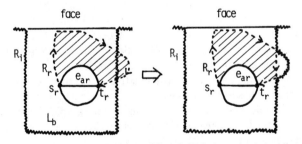

Fig. 11.16. Alternation of route R_r of flow f_r. (R_r is drawn by dashed lines.)

These two lemmas imply that MULTIFLOW2 correctly finds multicommodity flows satisfying given demands. However, concerning this point, it is not clear that MULTIFLOW2 terminates in polynomial-time or even finitely.

11.5.3. Refinement and complexity

In this subsection we first refine MULTIFLOW2 and then analyze the complexity.

Refinement. Although the given graph G is simple, G_a is not necessarily simple. Suppose that demand edge e_{ai} is parallel with an edge $e = (s_i, t_i) \in E$. Then $Q_{ij} = \{e\}$ where j is either 1 or 2, and hence $c_a(Q_{ij}) = \infty$ by the definition. Therefore $D = \min\{d_i, c(e)\}$, and hence e or e_{ai} is deleted from G_a when D units of f_i have been pushed through e. Repeating this procedure for all parallel edges, we will have a simple G_a.

We now modify MULTIFLOW slightly as follows: choosing Q_{ij}, one by one, in cyclic order, we check whether D of Q_{ij} is positive; and select the first one with positive D. Then Lemma 11.11 implies that we can find Q_{ij} with positive D among the first $2k$ paths. Furthermore we have the following lemma.

Lemma 11.12. *If G_a is a simple planar graph, then the sign of D is checked for at most $6kn$ paths in total during one execution of MULTIFLOW2.*

Proof. When $D(> 0)$ units of flow f_i have been pushed through path Q_{ij}, one of the following must occur:
(a) An edge of Q_{ij} is saturated and hence deleted;
(b) Flow f_i is satisfied, and hence demand edge e_{ai} is deleted; or
(c) The value D of Q_{ij} becomes zero.
Suppose that case (c) occurs. Then Q_{ij} now has two "blocking" edges e and e' such that $c_a(e, e') = 0$, that is, G_a has a cutset $E_a(X)$ such that $c_a(X) = 0$ and $E_a(X) \cap Q_{ij} = \{e, e'\}$. It is clear from the proof of Lemma 11.10 that $c_a(X)$ does not increase during the execution of MULTIFLOW2. Thus the value D of Q_{ij} will remain zero thereafter unless Q_{ij} is altered by the deletion of edges.

If either D was zero or case (c) occurred for each of $2k$ paths successively chosen, then G_a would have no Q_{ij} with positive D, contrary to Lemma 11.11. Therefore, among $2k$ successive trials, case (a) or (b) must occur, that is, at least one edge of G_a is deleted. Since G_a is a planar simple graph, G_a has at most $3n$ edges by Corollary 1.1. Hence the sign of D is checked for at most $6kn$ paths in total during one execution of MULTIFLOW2.   Q.E.D.

Lemma 11.12 implies that the sign of D is checked $O(kn)$ times in total during one execution of MULTIFLOW2, and hence MULTIFLOW2 runs in polynomial time. However a sophisticated method can improve the bound $O(kn)$ above into $O(n)$ (see [MNS86]).

The most time-consuming part of MULTIFLOW2 is the computation of D.

The problem is to compute $c_a(Q_{ij})$. Clearly $c_a(Q_{ij})$ is the length of a shortest cycle Z of G_a^* containing exactly two edges of Q_{ij}. Let v_0 be a vertex of G_a^* corresponding to the face of G_a bounded by $Q_{ij} \cup \{e_{ai}\}$. Let $H = G_a^* - e_{ai}^* = (U, F)$. Then Z is a shortest cycle in H passing through v_0.

We next show how to find Z. Replace every vertex of H except v_0 by a star graph, and replace v_0 by a graph obtained from a star graph by deleting edge (v_0', v_0''). (See Fig. 11.13.) Let H' be the resulting graph. Then one can easily obtain the following lemma.

Lemma 11.13. *If M is a complete matching of H', then $F - M$ is a union of vertex-disjoint cycles of H including a cycle passing through v_0. Moreover the converse is also true.*

Since G satisfies the cut condition, H has no negative cycle. Therefore, if M is a maximum weight complete matching of H', then $F - M$ consists of a shortest cycle of H passing through v_0 possibly together with vertex-disjoint zero cycles. Hence $c_a(Q_{ij})$ can be computed by finding a maximum weight complete matching of H'.

Thus algorithm MULTIFLOW2 finds k-commodity flows of given demands in the time required in order to find a maximum weight matching $O(n)$ times. Hence MULTIFLOW2 runs in $O(n^{5/2} \log n)$ time.

Theorem 11.6. *If network N belongs to class C_a, then MULTIFLOW2 finds k-commodity flows satisfying given demands in $O(n^{5/2} \log n)$ time, using $O(kn)$ space.*

REFERENCES

[AH77] K. Appel and W. Haken, Every planar map is four colourable, Part I: discharging, Illinois J. Math. 21 (1977) 429–490.

[AHU74] A. V. Aho, J. E. Hopcroft and J. D. Ullman, The Design and Analysis of Computer Algorithms (Addison-Wesley, Reading, Mass., 1974).

[AKS84] T. Asano, S. Kikuchi and N. Saito, A linear algorithm for finding Hamiltonian cycles in 4-connected maximal planar graphs, Discrete Appl. Math. 7 (1984) 1–15.

[Alb74] M. O. Albertson, Finding an independent set in a planar graph, in: Graphs and Combinatorics, eds., R. A. Bari and F. Harary (Springer-Verlag, Berlin, 1974) pp. 173–179.

[Alb76] M. O. Albertson, A lower bound for the independence number of a planar graph, J. Combinatorial Theory (B) 20 (1976) 84–93.

[AM70] J. G. Auguston and J. Minker, An analysis of some graph theoretical cluster techniques, J. Assoc. Comput. Mach. 17 (1970) 571–588.

[And77] L. D. Andersen, On edge-colouring of graphs, Math. Scand. 40 (1977) 161–175.

[ANW80] T. Asano, T. Nishizeki and T. Watanabe, An upper bound on the length of a Hamiltonian walk of a maximal planar graph, J. Graph Theory 4, 3 (1980) 315–336.

[AP61] L. Auslander and S. V. Parter, On imbedding graphs in plane, J. Math. Mech. 11, 3 (1961) 517–523.

[Arj82] E. Arjomandi, An efficient algorithm for coloring the edges of a graph with $\Delta + 1$ colours, INFOR 20, 2 (1982) 82–101.

[Bak83] B. S. Baker, Approximation algorithms for NP-complete problems on planar graphs, 24th Ann. Symp. on Found. of Comp. Sci. (1983) pp. 265–273.

[BD83] M. O. Ball and U. Derigs, An analysis of alternative strategies for implementing matching algorithms, Networks 13 (1983) 517–549.

[BE82] R. Bar-Yehuda and S. Even, On approximating a vertex cover for planar graphs, Proc. 14th Annual ACM Symp. on Theory of Computing, San Francisco, May 5–7 (1982) pp. 303–309.

[Ber57] C. Berge, Two theorems in graph theory, Proc. Nat. Acad. Sci. USA 43 (1957) 842–844.

[Ber75] J. C. Bermond, On Hamiltonian walks, in: Proc. 5th British Combinatorial Conf. Aberdeen (1975) pp. 41–51.

[Ber76] C. Berge, Graphs and Hypergraphs, 2nd edn. (North-Holland, Amsterdam, 1976).

[BL76] K. S. Booth and G. S. Lueker, Testing the consecutive ones property, interval graphs, and graph planarity using PQ-tree algorithms, J. Comput. Syst. Sci. 13 (1976) 335–379.

[BM76] J. A. Bondy and U. S. R. Murty, Graph Theory with Applications (Macmillan, London, 1976).

[Bol79] B. Bollobás, Graph Theory, An Introductory Course (Springer-Verlag, New York, 1979).

[Car79] B. Carre, Graphs and Network (Clarendon Press, Oxford, 1979).

[CH82] R. Cole and J. Hopcroft, On the coloring bipartite graphs, SIAM J. Comput. 11, 3 (1982) 540–546.

[Chv83] V. Chvátal, Linear Programming (W. H. Freeman and Company, New York, 1983).

[CL85] G. Chartrand and L. Lesniak, Graphs and Digraphs, 2nd edn (Warworth International, Belmont, California, 1985).

[CN84] N. Chiba and T. Nishizeki, The Hamiltonian cycle problem is linear-time solvable for 4-connected planar graph, TRECIS-8402, Tohoku Univ., Sendai, Japan (1984).

[CN85a] N. Chiba and T. Nishizeki, Arboricity and subgraph listing algorithms, SIAM J. Comput. 14, 1 (1985) 210–223.

[CN85b] N. Chiba and T. Nishizeki, A theorem on paths in planar graphs, J. of Graph Theory 10 (1985) 449–450.

[CNAO85] N. Chiba, T. Nishizeki, S. Abe and T. Ozawa, A linear algorithm for embedding planar graphs using PQ-trees, J. Comput. Syst. Sci. 30, 1 (1985) 54–76.

[CNS81a] N. Chiba, T. Nishizeki and N. Saito, Applications of the planar separator theorem, J. Information Processing 4, 4 (1981) 203–207.

[CNS81b] N. Chiba, T. Nishizeki and N. Saito, A linear 5-coloring algorithm of planar graphs, J. Algorithms 2 (1981) 317–327.

[CNS82] N. Chiba, T. Nishizeki and N. Saito, An approximation algorithm for the maximum independent set problem on planar graphs, SIAM J. Comput. 11, 4 (1982) 663–675.

[CNS83] N. Chiba, T. Nishizeki and N. Saito, An efficient algorithm for finding an independent set in planar graphs, Networks 13 (1983) 247–252.

[CON85] N. Chiba, K. Onoguchi and T. Nishizeki, Drawing plane graphs nicely, Acta Informatica 22 (1985) 187–201.

[Coo71] S. A. Cook, The complexity of theorem proving procedures, Proc. 3rd Ann. ACM Symp. on Theory of Computing (1971) 151–158.

[CR76] S. A. Cook and R. A. Reckhow, Time bounded random access machines, J. Comput. Syst. Sci. 7 (1976) 354–375.

[CYN84] N. Chiba, T. Yamanouchi and T. Nishizeki, Linear algorithms for convex drawings of planar graphs, in: Progress in Graph Theory, eds., J. A. Bondy and U. S. R. Murty (Academic Press, Toronto, 1984) pp. 153–173.

[Dd72] H. Diaz and G. de Ghellinck, Multicommodity maximum flow in planar networks (The D-algorithm approach), CORE discussion paper No. 7212, Center for Operations Research and Econometrics, Louvain-la-Neuve, Belgium (1972).

[Dij59] E. W. Dijkstra, A note on two problems in connexion with graphs, Numerische Mathematik 1 (1959) 269–271.

[EK75] S. Even and O. Kariv, An $O(n^{2.5})$ algorithm for maximum matching in graphs, Proc. 16th IEEE Symp. on FOCS (1975) pp. 100–112.

[ET76] S. Even and R. E. Tarjan, Computing an st-numbering, Theor. Comput. Sci. 2 (1976) 339–344.

[Eve79] S. Even, Graph Algorithms (Computer Science Press, Potomac, Maryland, 1979).

[Far48] I. Fary, On straight representations of planar graphs, Acta Sci. Math. Szeged 11 (1948) 229–233.

[FF56] L. R. Ford and D. R. Fulkerson, Maximal flow through a network, Canad. J. Math. 8 (1956) 399–404.

[Fio75] S. Fiorini, On the chromatic index of outerplanar graphs, J. Combinatorial Theory Ser. B 18 (1975) 35–38.

[Fre83] G. N. Frederickson, Shortest path problems in planar graphs (preliminary version), Proc. 24th Symp. on Foundation of Computer Science, Tucson, (1983) pp. 242–247.

[Fre84] G. N. Frederickson, On linear-time algorithms for five-coloring planar graphs, Information Processing Letters 19 (1984) 219–224.

[Fre85] G. N. Frederickson, Fast algorithms for shortest paths in planar graphs, with applications, CSD TR 486, Purdue University (1985).

[FW77] S. Fiorini and R. J. Wilson, Edge-Colourings of Graphs (Pitman, London, 1977).

[Gal83] Z. Galil, Efficient algorithms for finding maximal matching in graphs, Computing Surveys 18, 1 (1986) 23–38.

[Gar76] M. R. Garey, D. S. Johnson and R. E. Tarjan, The planar Hamiltonian circuit problem is NP-complete, SIAM J. Comput. 5 (1976) 704–714.

[GH74] S. E. Goodman and S. T. Hedetniemi, On Hamiltonian walks of graphs, SIAM J. Comput. 3 (1974) 214–221.

[GJ76] M. R. Garey and D. S. Johnson, The complexity of near-optimal graph coloring, J. Assoc. Comput. Mach. 23 (1976) 43–49.

[GJ79] M. R. Garey and D. S. Johnson, Computers and Intractability (W. H. Freeman and Company, San Francisco, 1979).

[GJS76] M. R. Garey, D. S. Johnson and L. Stockmeyer, Some simplified NP-complete graph problems, Theor. Comput. Sci. 1 (1976) 237–267.

[GK82] H. N. Gabow and O. Kariv, Algorithms for edge coloring bipartite graphs and multigraphs, SIAM J. Comput. 11, 1 (1982) 117–129.

[GMG82] Z. Galil, S. Micali and H. Gabow, Priority queues with variable priority and an $O(EV \log V)$ algorithm for finding a maximal weighted matching in general graphs, 23rd Ann. Symp. on Found. of Comput. Sci., Chicago (1982) pp. 255–261.

[GNKLT84] H. N. Gabow, T. Nishizeki, O. Kariv, D. Leven and O. Terada, Algorithms for edge-coloring graphs, submitted to a journal (1984).

[Gol63] A. J. Goldstein, An efficient and constructive algorithm for testing whether a graph can be embedded in a plane, Graph and Combinatorics Conf., Contract No. NONR 1858–(21), Office of Naval Research Logistics Proj., Dept. of Math., Princeton Univ. (1963).

[Gol73] M. K. Goldberg, On multigraphs with almost maximal chromatic class (in Russian), Diskret Analiz 23 (1973) 3–7.

[Gol84a] M. K. Goldberg, Edge-coloring of multigraphs; Recoloring technique, J. Graph Theory 8 (1984) 123–137.

[Gol84b] M. Goldberg, An approximate algorithm for the edge-coloring problem, Congressus Numerantium 43 (1984) 317–319.

[Gou82] D. Gouyou-Beauchamps, The Hamiltonian circuit problem is polynomial for 4-connected planar graphs, SIAM J. Comput. 11 (1982) 529–539.

[Har72] F. Harary, Graph Theory (Addison-Wesley, Reading, Mass., 1972) (revised).

[Has81] R. Hassin, Maximum flow in (s, t) planar networks, Inf. Proc. Lett. 13, 3 (1981) 107.

[Has84] R. Hassin, On multicommodity flows in planar graphs, Networks 14 (1984) 225–235.

[HJ85] R. Hassin and D. B. Johnson, An $O(n \log^2 n)$ algorithm for maximum flow in undirected planar networks, SIAM J. Comput. 14, 3 (1985) 612–624.

[HNS84] D. S. Hochbaum, T. Nishizeki and D. S. Shmoys, A better than "best possible" algorithm to edge color multigraphs, J. Algorithms 7 (1986) 79–104.

[Hol81] I. J. Holyer, The NP-completeness of edge colourings, SIAM J. Comput. 10 (1981) 718–720.

[HT73] J. E. Hopcroft and R. E. Tarjan, Dividing a graph into triconnected components, SIAM J. Comput 2, 3 (1973) 135–158.

[HT74] J. E. Hopcroft and R. E. Tarjan, Efficient planarity testing, J. Assoc. Comput. Mach. 21 (1974) 549–568.

[Hu69] T. C. Hu, Integer Programming and Network Flows (Addison-Wesley, Reading, Mass., 1969).

[HU79] J. E. Hopcroft and J. D. Ullman, Introduction to Automata Theory, Languages, and Computation (Addison-Wesley, Reading, Mass., 1979).

[IR78] A. Itai and M. Rodeh, Finding a minimum circuit in a graph, SIAM J. Comput. 7, 4 (1978) 413–423.

[IS79] A. Itai and Y. Shiloach, Maximum flows in planar networks, SIAM J. Comput. 8, 2 (1979) 135–150.

[Ita78] A. Itai, Two-commodity flow, J. Assoc. Comput. Mach. 25, 4 (1978) 596–611.

[Joh77] D. B. Johnson, Efficient algorithms for shortest paths in sparse networks, J. Assoc. Comput. Mach. 24 (1977) 1–13.

[Jol75] J. L. Jolivet, Hamiltonian pseudo cycles in graphs, Proc. 5th Southeastern Conference Combinatorics, Graph Theory and Computing, Boca Raton (1975) pp. 529–533.

[JV82] D. B. Johnson and S. M. Venkatesan, Using divide and conquer to find flows in directed planar networks in $O(n^{3/2} \log n)$ time, Proc. 20th Ann. Allerton Conf. on Communication, Control, and Computing, Univ. of Illinois (1982) pp. 898–905.

[Kar72] R. M. Karp, Reducibility among combinatorial problems, in: Complexity of Computer Computations, eds. R. E. Miller and J. W. Thatcher (Plenum Press, New York, 1972) pp. 85–104.

[Kar84] N. Karmarkar, A new polynomial-time algorithm for linear programming, Proc. 16th Ann. ACM Symp. on Theory of Computing (1984) pp. 302–311.

[KD79] M. S. Krishnamoorthy and N. Deo, Node-deletion *NP*-complete problems, SIAM J. Comput. 8, 4 (1979) 619–625.

[Kha79] L. G. Khachiyan, A polynomial algorithm in linear programming, Dokl. Akad. Nauk SSSR N. S. 244 : 5 (1979) 1093–1096. [English transl., Soviet Math. Dokl. 20 : 1 (1979) 191–194.]

[Law76] E. L. Lawler, Combinatorial Optimization: Networks and Matroids (Holt, Rinehart and Winston, New York, 1976).

[LEC67] A. Lempel, S. Even and I. Cederbaum, An algorithm for planarity testing of graphs, Theory of Graphs, Int. Symp. Rome, July, 1966, ed., P. Rosenstiehl, (Gordon and Breach, New York, 1967) pp. 215–232.

[Lei80] C. E. Leiserson, Area-efficient graph layouts (for VLSI), Carnegie-Mellon University, CMU-CS-80-138 (1978).

[LM78] R. J. Lipton and R. E. Miller, A batching method for coloring planar graphs, Information Processing Letters 7, 4 (1978) 185–188.

[LRT79] R. J. Lipton, D. J. Rose and R. E. Tarjan, Generalized nested dissection, SIAM J. Numer. Anal. 16, 2 (1979) 346–358.

[LT79] R. J. Lipton and R. E. Tarjan, A separator theorem for planar graphs, SIAM J. Appl. Math. 36 (1979) 177–189.

[LT80] R. J. Lipton and R. E. Tarjan, Applications of a planar separator theorem, SIAM J. Comput. 9, 3 (1980) 615–627.

[Mac37] S. MacLane, A combinatorial condition for planar graphs, Fund. Math. 28 (1937) 22–32.

[Mil84] G. Miller, Finding small simple cycle separators for 2-connected planar graphs, Proc. 16th Ann. ACM Symp. on Theory of Computing (1984) pp. 376–382.

[MNS85] K. Matsumoto, T. Nishizeki and N. Saito, An efficient algorithm for finding multicommodity flows in planar networks, SIAM J. Comput. 14 (1985) 289–301.

[MNS86] K. Matsumoto, T. Nishizeki and N. Saito, Planar multicommodity flows, maximum matchings and negative cycles, SIAM J. Comput. 15, 2 (1986) 495–510.

[MST80] D. W. Matula, Y. Shiloach and R. E. Tarjan, Two linear-time algorithms for five-coloring a planar graph, Manuscript (1980).

[MV80] S. Micali and V. V. Vazirani, An $O(|V|^{1/2}|E|)$ algorithm for finding maximum matching in general graphs, 21th Ann. Symp. on Found. of Comp. Sci. (1980) pp. 17–27.

[Nas61] C. St. J. A. Nash-Williams, Edge-disjoint spanning trees of finite graphs, J. London Math. Soc. 36 (1961) 445–450.

[NAW83] T. Nishizeki, T. Asano and T. Watanabe, An approximation algorithm for the Hamiltonian walk problem on a maximal planar graph, Discrete Applied Math. 5 (1983) 211–222.

[NB79] T. Nishizeki and I. Baybars, Lower bounds on the cardinality of the maximum matchings of planar graphs, Discrete Math. 28 (1979) 255–267.

[Nis79] T. Nishizeki, On the relationship between the genus and the cardinality of the maximum matchings of a graph, Discrete Math. 25, 2 (1979) 149–156.

[NK84] T. Nishizeki and K. Kashiwagi, On the 1.1 edge-coloring of multigraphs, TRECIS-84003, Tohoku Univ., Sendai, Japan (1984).

[NS84] T. Nishizeki and M. Sato, An algorithm for edge-coloring multigraphs (in Japanese), Inst. Elect. Commun. Eng, Trans. J67-D, 4 (1984) 466–471.

[NSS85] T. Nishizeki, N. Saito and K. Suzuki, A linear-time routing algorithm for convex grids, IEEE Trans. on Computer-Aided Design, CAD-4, 1 (1985) 68–76.

[Oka83] H. Okamura, Multicommodity flows in graphs, Discrete Appl. Math. 6 (1983) 55–62.

[Ore62] O. Ore, Theory of Graphs, Amer. Math. Soc. Colloq. Publ., 38, Providence, R.I. (1962).

[Ore67] O. Ore, The Four Color Problem (Academic Press, New York, 1967).

[OS81] H. Okamura and P. D. Seymour, Multicommodity flows in planar graphs, J. Combinat. Theory Series B31 (1981) 75–81.

[PY81] C. H. Papadimitriou and M. Yannakakis, The clique problem for planar graphs, Information Processing Letters 13, 4, 5 (1981) 131–133.

[Rei83] J. H. Reif, Minimum $s - t$ cut of a planar undirected network in $O(n \log {}^2(n))$ time, SIAM J. Comput. 12 (1983) 71–81.

[Ric86] D. Richards, Finding short cycles in planar graphs using separators, J. Algorithms 7 (1986) 382–394.

[RT75] R. C. Read and R. E. Tarjan, Bounds on backtrack algorithms for listing cycles, paths, and spanning trees, Networks 5 (1975) 237–252.

[RT81] E. M. Reingold and J. S. Tilford, Tidier drawings of trees, IEEE Trans. Software Engineering 7 (1981) 223–228.

[Sak66] M. Sakarovitch, The Multicommodity Flow Problem, Doctoral thesis, Oper. Res. Center, University of California, Berkeley (1966).

[Sak73] M. Sakarovitch, Two commodity network flows and linear programming, Math. Prog. 4 (1973) 1–20.

[SDK83] M. M. Sysło, N. Deo and J. S. Kowalik, Discrete Optimization Algorithms (Prentice-Hall, Englewood Cliffs, NJ, 1983).

[Sey79] P. D. Seymour, A short proof of the two-commodity flow theorem, J. Comb. Theory, Series B26 (1979) 370–371.

[Sey81] P. D. Seymour, On odd cuts and planar multicommodity flows, Proc. London Math. Soc. (3) 42 (1981) 178–192.

[Sha49] C. E. Shannon, A theorem on colouring the lines of a network, J. Math. Phys. 28 (1949) 148–151.

[Shi79] Y. Shiloach, Union-member algorithms for non-disjoint sets, Technical Report STAN-CS-728, Computer Science Department, Stanford University (1979).

[SK77] T. L. Saaty and P. C. Kainen, The Four-Colour Problem (McGraw-Hill, New York, 1977).

[Sle80] D. D. Sleator, An $O(nm \log n)$ Algorithm for Maximum Network Flow, PhD Dissertation, Computer Science Department, Stanford University, Stanford, California (1980).

[SNS85] H. Suzuki, T. Nishizeki and N. Saito, Multicommodity flows in planar undirected graphs and shortest paths, Proc. 17th Ann. ACM Symp. on Theory of Computing, Providence, RI (1985) pp. 195–204.

[SR83] K. J. Supowit and E. M. Reingold, The complexity of drawing trees nicely, Acta Informatica 18 (1983) 377–392.

[ST83] D. D. Sleator and R. E. Tarjan, A data structure for dynamic trees, J. Comput. Syst. Sci. 26 (1983) 362–390.

[Tan64] D. T. Tang, Bi-path networks and multicommodity flows, IEEE Trans. Circuit Theory CT-11 (1964) 468–474.

[Tho80] C. Thomassen, Planarity and duality of finite and infinite graphs, J. Combinat. Theory, Series B29 (1980) 244–271.

[Tho83] C. Thomassen, A theorem on paths in planar graphs, J. Graph Theory 7 (1983) 169–176.

[TIAS77] S. Tsukiyama, M. Ide, H. Ariyoshi and I. Shirakawa, A new algorithm for generating all the maximal independent sets, SIAM J. Comput. 6, 3 (1977) 505–517.

[TN82] O. Terada and T. Nishizeki, Approximate algorithms for the edge-coloring of graphs (in Japanese), Trans. Inst. of Elect. and Commun. Eng. of Japan, J65-D, 11 (1982) 1382–1389.

[Tob75] R. L. Tobin, Minimal complete matchings and negative cycles, Networks 5 (1975) 371–387.

[Tur36] A. M. Turing, On computable numbers, with an application to the Entscheidungs problem, Proc. London. Math. Soc., 2–42 (1936) pp. 230–265; Correction, ibid, 2–43, pp. 544–546.

[Tut47] W. T. Tutte, The factorization of linear graphs, J. London Math. Soc. 22 (1947) 107–111.

[Tut54] W. T. Tutte, A short proof of the factor theorem for finite graphs, Canad. J. Math. 6 (1954) 347–352.

[Tut56] W. T. Tutte, A theorem on planar graphs, Trans. Amer. Math. Soc. 82 (1956) 99–116.

[Tut60] W. T. Tutte, Convex representations of graphs, Proc. London Math. Soc. (3) 10 (1960) 304–320.

[Tut63] W. T. Tutte, How to draw a graph, Proc. London Math. Soc. 13 (1963) 743–768.

[Tut77] W. T. Tutte, Bridges and Hamiltonian circuits in planar graphs, Aequationes Mathematica 15 (1977) 1–33.

[Vau80] J. G. Vaucher, Pretty-printing of trees, Software-Practice and Experience 10 (1980) 553–561.

[Viz64] V. G. Vizing, On an estimate of the chromatic class of a p-graph (in Russian), Diskret Analiz 3 (1964) 23–30.

[Viz65a] V. G. Vizing, Critical graphs with a given chromatic class (in Russian), Diskret Analiz 5 (1965) 9–17.

[Vis65b] V. G. Vizing, The chromatic class of a multigraph, Cybernetics 3 (1965) 32–41.

[Yan78] M. Yannakakis, Node- and edge-deletion NP-complete problems, Proc. 10th ACM Symp. on Theory of Computing (1978) 253–264.

[Yap81] H. P. Yap, On graphs critical with respect to edge-colourings, Discrete Math. 37 (1981) 289–296.

[Whi33a] H. Whitney, Planar graphs, Fund. Math. 20 (1933) 73–84.

[Whi33b] H. Whitney, 2-isomorphic graphs, Amer. J. Math. 55 (1933) 245–254.

[Wig82] A. Wigderson, A new approximate graph coloring algorithm, Proc. 14th Ann. ACM Symp. on Theory of Computing, San Francisco, May 5–7 (1982) pp. 325–329.

[Wig83] A. Wigderson, Improving the performance guarantee for approximate graph coloring, J. Assoc. Comput. Mach. 30, 4 (1983) 729–735.

[Wil85] R. J. Wilson, Introduction to Graph Theory, 3rd ed. (Longman, London, 1985).

[WS70] C. Wetherell and A. Shannon, Tidy drawings of trees, IEEE Trans. Software Engineering 5 (1970) 514–520.

INDEX

227